PORT PHILLIP:
PILOTS AND DEFENCES

By the same author

Australian Lighthouses

Hazards of the Sea

PORT PHILLIP: PILOTS AND DEFENCES

BY

CAPTAIN J. NOBLE

MELBOURNE

THE HAWTHORN PRESS

First Published 1973

SBN 7256 0111 6

Registered at the GPO Melbourne
for transmission by post as a book

Wholly set up and printed in Australia by
The Hawthorn Press Pty Ltd
601 Little Bourke Street, Melbourne 3000

To all ex-pilots of the Service,
wherever they may be,
who have left as their memorial
an enduring name for personal uprightness
and fearless devotion to duty

Contents

Illustrations

Acknowledgments

Research for this segment of the maritime history of Port Phillip extends over several years and material has been obtained from several sources, none of which is of greater value than the other.

I would like to record my grateful appreciation for the assistance received from the staff of the La Trobe Library; to the Royal Historical Society of Victoria in general and to Mrs Maxwell, Mrs Hosken and Mrs Hill in particular. Also to Mr John Ware and his assistants in the historical records section of Victoria Barracks; to Mr J. Wagglen, Director of the Ports and Harbours Department, for access to their files, and to Garry Tomlin for his assistance with those files.

Jack Gange of the Queenscliff Historical Society, and Ned Stonehouse of Point Lonsdale lighthouse have also rendered valuable assistance in the research and with illustrations.

Appreciation also to The Herald and Weekly Times for permission to use illustrations from their files.

Many thanks, also to former pilot Captain George Molyneux who provided the nucleus which resulted in this history of the Port Phillip Pilots.

Acknowledgements

Foreword

It is with great pleasure that I write this Foreword to Captain John Noble's comprehensive book on the colourful history of the Port Phillip Pilot Service.

Its publication coincides with the centenary of the *Rip* disaster when four lives were lost. This tragic happening – on 15 July 1873 – was, as Captain Noble states 'the worst ever to befall the Port Phillip Pilot Service'.

Captain Noble's account of the tragedy, and of the commercial ships involved in disaster or near disaster in and out of Port Phillip, is most graphic and provides fascinating reading.

His recording of the role of pilots since the early days of the 19th century, and the founding of the present service, is the work of a dedicated man and reflects long research and study of this vital chapter in Victoria's maritime history.

It is also a moving tribute to the selfless bravery and devotion to duty of this fine body of men over many years.

Captain Noble is to be commended for producing this fine book, which I am sure will be of great interest to all sections of the public as a record of the contribution made by the Pilot Service to the development of this great State.

R. J. HAMER

PART ONE

Port Phillip: Pilots

CHAPTER 1

Times and Tides

All early explorers who entered Port Phillip were impressed by the extreme turbulence of the mile-and-three-quarter wide entrance, particularly when the outgoing tide met ocean swells created by persistent onshore winds. Masters of sailing ships making a landfall at Port Phillip after the long haul across the Southern Ocean found themselves with no alternative to standing on regardless of turbulence or tide when the winds blew onshore.

Tidal streams flow for six and a quarter hours from low water to high water and then ebb for a similar period. Obviously the most suitable time for a sailing ship to arrive off the Heads would be at the slack water period at the end of the ebb and the beginning of the flood stream. But the tidal streams of the Rip are at their strongest at the predicted times of high and low water at Port Phillip Heads. This is because the tidal range in Bass Strait is eight feet. While the tide is rising in the Strait the narrow entrance to Port Phillip only permits sufficient water to enter to raise the level inside the seven hundred and twenty-eight square mile bay by between two and three feet, which is the tidal range for all places within Port Phillip. Therefore, when the tide is high in Bass Strait – the time of high water at Port Phillip Heads – the sea level is about three feet lower inside the Heads and the flood stream will continue for another three hours until both levels are the same.

In Bass Strait the tide will then have been falling for three hours, and will continue to fall for another three hours at a greater rate than it does inside. At low water in Bass Strait the levels are about three feet lower and the outgoing, or ebb, stream continues until both levels equalise again.

Shipmasters bound for Port Phillip, aware of the predicted times for high and low water at the Heads, often failed to appreciate that the times of slack water occurred three hours later than these predicted times. Arriving at a time they expected slack water the streams would be at their maximum – varying between seven and twelve knots and potentially dangerous for vessels depending upon sails for propulsion.

Between Port Phillip Heads the deepest water extends in a rough semi-circle around Point Nepean. Immediately seawards of this gutter, athwart of the fairway, is Rip Bank – a ledge of sandstone which restricts the draught of vessels using Port Phillip. Rip Bank deflects the southwesterly flow of the ebb stream into a southeasterly direction, creates the notorious turbulence of the Rip, and sets approaching ships towards the dangerous rocks near Point Nepean. Shipmasters without local knowledge, who timed their arrival for what they expected would be the beginning of the ingoing stream, suddenly found their ships unmanageable in the swirling maelstrom of a maximum ebb stream. This combination of circumstances accounts for many of the long list of wrecks near Port Phillip Heads.

Inquiries into these wrecks revealed that a recurring cause was captains attempting to enter at night against an ebb tide with insufficient local knowledge. Encouraged by a fair wind offshore their ships were swept towards Point Nepean and were in trouble before the unwary captains appreciated the fact.

Corsair Rock is the greatest danger. Although its approximate position was known to pilots and local shipmasters before it was given this name on 19 October 1853, after the harbour master, Captain Ferguson, instructed the pilots in charge of the *Anonyma* and *Corsair* to take careful soundings while the tide was slack. Protruding into the fairway from Point Nepean there is only eight feet of water over Corsair Rock. Knowledge of its exact position did not bring to an end the long list of ships striking the rock.

Half a mile northeastwards of Corsair Rock is a rocky islet between Point Nepean and Nepean Reef. Beacons on this islet,

Time-ball tower lighthouse today, the base is the original Gellibrand Point lighthouse

Shipping anchored in Hobson's Bay 1853, with the original Gellibrand Point lighthouse in the foreground

The pilot steamer Victoria, 1901-1957

The former steam yacht Alvina, used as a Port Phillip pilot steamer 1903-1925

known as Beacon Rock, and on Point Nepean, when in transit lead over Corsair Rock on the eastern side and over Lonsdale Rock on the western side, of the fairway. Lonsdale Rock, with twenty-two feet of water over it, lies half a mile southeastwards from Point Lonsdale. Inside Lonsdale Rock is a channel for light draught vessels which avoids the full strength of the tide and the extreme turbulence of the Rip.

When the famous clipper ship *Lightning* sailed for England in 1863 with a draught of nineteen feet an underwater obstruction was encountered about a mile south of Point Nepean. Investigations revealed the ship was not making water and Captain Clark elected to continue his voyage. *Lightning* was fortunate not to have joined the long list of ships posted missing on similar voyages; because, when drydocked at Liverpool, a piece of sandstone from the seabed near the Rip was found securely lodged in a jagged hole in her hull.

Immediately this information reached Melbourne Captain Ferguson ordered an extensive survey of the area where the obstruction was encountered. This revealed a rocky shoal in the fairway carrying only eighteen feet of water. Although £1000 was immediately allotted for the removal of Lightning Rocks the danger existed until 1883, when a series of blasting operations levelled the pinnacles to a depth of thirty-four feet.

In June 1841 Shortland's Bluff was selected as the most suitable position for a lighthouse, one reason being that a light here would guide ships along the channels as well as through the entrance. The lighthouse was completed in 1843 at a cost of £1900 – a forty-five foot tower built of local sandstone and containing a lantern with eight oilburners in a circular column to project the main light, with an additional burner adjusted to provide a beam of light through the entrance fairway.

Seven years later another light, shown from a small octagonal red gum tower, known as the Low Light, was established eight hundred feet southwest of the Shortland's Bluff lighthouse. Both lights in line led ships midway between Points Lonsdale and Nepean. By 1856 the local sandstone in the Shortland's

Bluff lighthouse was eroding and plans were approved for new and improved leading lights. British stonemasons had become highly specialised at building lighthouses. Blocks of stone were shaped and assembled in the required form and carefully marked before being dismantled and shipped to the selected site for re-assembly. This is how the existing eighty-one foot bluestone tower on Shortland's Bluff came into being. Because of its retained natural colour it has always been known as the Black Light. Holes made for lifting each block into position are still clearly visible. The tower was completed in 1862 in a position behind, but on the same bearing as the original lighthouse, which was then demolished. The following year a lighthouse seventy-three feet high and painted white, was erected one thousand and sixty-six feet in front of the Black light and on the same bearing as the octagonal tower.

This red gum structure, still in excellent condition, was taken to Point Lonsdale where Captain Preston, the signal master, had kept an ineffective oil light burning from his flagstaff since 1853. Painted red and white the octagonal tower housed a ten mile radius light until the present Point Lonsdale lighthouse was built in 1902.

CHAPTER 2

The First Pilots

The need for pilotage assistance at Port Phillip was apparent from the time commercial ships began using the port. In addition to the entrance dangers there are long channels to traverse, with varying depths. Both Melbourne and Geelong have always been restricted by navigational hazards that necessitated local knowledge to ensure shipping a safe passage.

Earliest pilots operated from Shortland's Bluff, using whaleboats to board incoming ships. They could not always proceed outside the Heads to assist ships past the dangers in the approaches. One of the earliest losses attributed to lack of pilotage assistance was the wreck of the *William Salthouse* late in 1841.

This three hundred and six ton barque brought the first cargo direct from British North America to Port Phillip – including flour, pork, dried fish, vinegar, cider and timber – and arrived from Quebec in stormy weather during the afternoon of 27 November. Trapped in the unexpected southeasterly set of the ebb tide she struck the outer edge of the reef (probably Corsair Rock) and became unmanageable when the rudder was damaged. When the flood tide made she was carried inside and beached on Swan Spit to prevent her from sinking. Three days later the *William Salthouse* was refloated, only to founder in three and a quarter fathoms of water on Popes Eye Shoal.

Pilots had then been operating in a private capacity for about three years. George Tobin is reputed to have performed the first pilotage duty when he brought the schooner *Industry* from Launceston to Melbourne, drawing eight and a half feet. Captain James Liddell, owner of the *Industry*, made his first appearance at Launceston as sailing master of the *Jolly Rambler* in 1832. Two years later he was in command of Henty's *Thistle*. After returning from a voyage to New Zealand in the *Thistle* it was Liddell who recommended Portland to the Henty family as a desirable place for a settlement on the mainland.

Captain Liddell later bought the *Industry* and chartered her to John Pascoe Fawkner to carry supplies across Bass Strait. He also used the schooner for whaling between Wilson's Promontory and Cape Schanck. After a spell in New Zealand he returned to Melbourne in the *Jenny Lind*.

At Williamstown on 18 January 1853 Captain Liddell of the *Jenny Lind*, appeared before the Magistrates Court for 'not paying 5 shillings per adult passenger under the Passenger Act'. Apparently Liddell did not appear. The *Herald* reported that 'he had been stuck up in Melbourne and his corporation so materially damaged as to confine him to his bed'. He was admonished for the offence *in absentia*.

Not long afterwards he was employed in the Port Phillip lighthouse service from which he retired on 3 February 1869 when he was master of the West Channel lightship.

George Tobin was reputed to have arrived at Port Phillip from Van Diemen's Land in James Simpson's eighteen ton sloop *Childe Harold* on 3 March 1837, bringing cargo of eight hundred bricks. Subsequently the *Childe Harold* traded within Port Phillip and made occasional voyages to Westernport. When anchored off the Werribee River on 19 March 1838 the *Childe Harold* was seized by bushrangers, who had also stolen goods from Simpson's store. After being chased by the revenue cutter *Ranger* the fugitives beached the *Childe Harold* near the Heads.

The *Childe Harold* ended her days, wrecked on Gellibrand's Point in a southerly gale in September 1841. Tobin had been in command of the sloop for two and a half years when he applied for a pilot's licence in March 1839. He was recommended by a group of Melbourne merchants, traders and shipmasters as 'a seaman and shipmaster for nearly twenty years who had gained an intimate knowledge of all the channels, soundings, etc, in the bay as well as in the Yarra Yarra river'.

Evidence was produced that Tobin had also sailed as master in the West Indian trade. His age was given as thirty-two. He was five feet five inches in height, of ruddy complexion and his place of abode Williamstown. At the first land sale he is recorded as having bought a block of land in Queen Street, Melbourne. He was described as a 'sober, steady and industrious man well qualified and fully competent in every respect to hold the situation of pilot'.

Twenty-three 'citizens, traders and shipmasters' signed a petition which was sent direct to the governor of New South Wales, Sir George Gipps; who replied that it should have been sent through the Port Phillip Police Magistrate, Captain Lonsdale. However, he recommended that the Sydney harbour master consult Captain Wickham on the necessity for pilots at Port Phillip.

Captain Wickham and the harbour master agreed that more than one pilot was necessary because of the increasing importance of the Port Phillip settlement, the number of vessels calling there, and 'it often appears that vessels take the ground which would be avoided by having some pilots acquainted with the navigation of this harbour and coast'. Tobin's application was then approved and it was suggested that the same pilotage rates as applied at Port Jackson be charged, but that the government 'not be put to any expense whatsoever'. Tobin's licence was sent to Melbourne on 26 June 1839.

Within four months another petition was presented on behalf of Thomas Halk Sutton 'praying for a pilot's licence for the River Yarra and the south and west entrance of the bay of Port Phillip'. Captain Sutton, aged forty-five, five foot six inches tall, a native of Kent residing in Queen Street, Melbourne, was a master mariner who owned and operated the *Letitia,* 'a fine schooner of twenty tons which will enable him to proceed to the Heads whenever a case of pilotage is required'.

Twenty-six 'merchants traders and master mariners' signed the petition. Sutton's licence was issued on 26 November. William Timothy was the next pilot appointed; followed during the next two years by Alexander McPherson and Thomas McClelland. Another pilot named Trundle was appointed in 1841 but there

WILLIAMSTOWN.

Pilot Board Offices,

________________187

SIR,

You having been appointed to take charge of______________

____________________ you are directed to be on

board at ____________________________

____

SECRETARY PILOT BOARD.

To Mr. Pilot____________________

is no record of his service after the harbour master reported to Superintendent La Trobe that Trundle had boarded the *Georgiana* 'in a tipsey condition and had abused the captain and the harbour master'.

Captain Charles Morgan Lewis was Port Phillip's first harbour master, taking up his duties on 21 October 1839 at a salary of £200 per year. Three years previously Captain Lewis was in command of the New South Wales government schooner *Isabella* on a voyage from Sydney to rescue survivors from the barque *Charles Eaton,* wrecked in the Torres Strait area in 1834. Some time after the wreck five survivors landed at Amboyna but the fate of twenty-seven others was not determined until after the ship *Mangles* arrived at Sydney and reported a white youth living with natives on Murray Island. The *Isabella* investigated.

Captain Lewis found John Ireland, youngest member of the crew of the *Charles Eaton,* and William D'Oyley, infant son of passengers on the barque. Ireland told of the massacre of twenty-five other survivors from the wreck, and of his and William's adoption by the natives. Captain Lewis then located the island where the massacre occurred, collected forty-five human skulls as evidence, and sailed for Sydney via Batavia and Western Australia. He later accompanied D'Oyley to England and delivered him into the care of relatives.

In London in March 1839 Captain Lewis successfully applied to the Secretary of State for Colonies, Lord Normanby, for the position of harbour master at Port Phillip. Although he commenced duties when he returned to Melbourne the appointment was not officially confirmed until 3 May 1841, by which time his salary had been increased from £200 to £250 per annum.

One of the new harbour master's first recommendations was that a pilot vessel cruise outside Port Phillip Heads. Captain Lewis did not consider whaleboats sufficiently seaworthy to work in the Rip. These pilot boats were about thirty feet long, double-ended, and handled by a coxswain with a steering oar and four oarsmen; and sails could be used effectively. He also recommended that two more pilots be appointed: Captains D. Morris and A. McPherson. The latter obtained his licence on 6 August 1840, but disaster overtook Morris before his licence was granted. With another prospective pilot named Delaney he sailed from Williamstown in mid July, in a boat bought from the captain of the *Duchess of Kent,* to meet an inward bound ship. This boat was unsuitable for Port Phillip's winter westerlies and nothing more was heard of the two prospective pilots. Wreckage of the boat, which clearly indicated it had capsized in a squall, was later found twenty miles south of Red Bluff.

Early in 1840 the harbour master suggested that a five-man Pilot Board be established to decide the 'important and difficult affair' of determining pilotage fees. He prepared a table of fees for the guidance of the board, on which were two government and two merchant representatives and the harbour master himself. Fees suggested for vessels of fifty to one hundred tons, not colonially registered, were £3-10, increasing to £6-16 for vessels of eight hundred and fifty tons, the largest visiting the port at that time. The Board was also required to report on the application of the Sydney Port regulations to Port Phillip, the pilotage system, and also the adoption of the Sydney quarantine regulations.

Their report suggested that Sydney's port regulations needed revising if they were to be applied to Port Phillip; that the scale of pilotage fees in Sydney were reasonable for this port except that 'frequent gales of wind and strong tide ripplings which persist in the bay at all periods of the year' made it impossible for open boats to be used. Nor did the Board consider that Port Phillip pilots could make a living without some additional remuneration.

It was Captain Lewis who first suggested a cruising cutter of thirty to thirty-five tons to maintain a cruising pilot station outside Port Phillip Heads; the vessel to be 'properly provisioned and kept in repair by the pilots, who would contribute a percentage of their fees to defray the cost'. Other Board members stipulated that the pilot ship should be the usual residence of the pilots who would no longer be permitted to erect huts or cottages at Shortland's Bluff on the grounds that 'if once domesticated they might feel unwilling to put to sea in bad weather'.

While these discussions took place in Melbourne the Port Phillip pilots were also under scrutiny in Sydney. On 8 August 1840 Captain Nicholson, the Sydney harbour master, reported to Governor Gipps that 'he hoped soon to have Port Phillip harbour and pilots on a better footing'; recommending two whaleboats be supplied to the four pilots and four crown prisoners to assist the four boatmen. One whaleboat, two pilots and a crew were to be always available.

The forty-six ton revenue cutter *Ranger*, due to return from New Zealand, could be converted into a cruising pilot vessel at small cost and maintained for about £400 a year. Captain Nicholson was also of the opinion that, considering the severity of the seas, Port Phillip pilots should be paid more than the Sydney pilots, and that twenty-five per cent be added to the Port Phillip rates to meet the cost of the *Ranger*. This vessel, which was built on Brisbane Water in 1837, with two whaleboats on addition to her own boats, was handed over to Captain Lewis on 21 April 1841. It was stipulated that not more than two pilots and one whaleboat were to leave the *Ranger* at any time other than on duty. The cutter's captain was to also act as customs officer and would be paid by the Customs Department for this.

Meanwhile the fortunes of the Port Phillip pilots fluctuated. Accounts for 1840 show that they employed five men, each at a salary of £1 per week. Provisions for these men cost 17s a week and £27 was spent on 'furnishing tents for pilots and crew'. After deducting boat hire at Williamstown there was only £104 to be divided among the four pilots for three months work. Cost of two whaleboats, £80, was not shown in the accounts.

In March 1841 the pilots complained to Superintendent La Trobe that their remuneration was insufficient to cover expenses and requested government assistance. The *Herald* newspaper, in an editorial, supported their claim:

> By reasonable exertions the Sydney pilots in the mere discharge of their duties netted a handsome income which by no means is too much for their fatigue, exertion, and danger they encounter.

> But at Port Phillip the dangers are far greater, and it is not possible for them to earn a decent livelihood because, to begin with, they have to provide a boat and crew – at least four men at £4 a head – also each man has to construct his own residence. Moreover the trip up the bay is so long that no more than one trip a day can be made and they may have to wait for weeks for another vessel. In bad weather, even in the Bay, an open boat cannot live, and so how impossible is it to survive in the open sea.
>
> At Sydney a pilot may take several vessels to an anchorage in a day and so a pilot's income is in the vicinity of £600 a year, whereas the Victorian pilots can hardly earn from £8 to £10 a month. We suggest that boats and crews be supplied to them which would mean a saving of about £400 a year.

Superintendent La Trobe reacted favourably. The *Ranger*, with two extra boats went into service. At the close of 1841 there were six pilots at Shortland's Bluff. Pilotage fees for the year amounted to £3019, one-sixth of which went towards the *Ranger's* expenses, leaving £2591 to be divided among the six pilots.

This improvement in the pilots' status prompted Superintendent La Trobe to recommend that pilotage fees be reduced and that vessels trading to Van Diemen's Land and New South Wales be charged half rates. He also suggested that pilots remuneration be fixed at £300 a year. Pilotage fees were then paid to the harbour master whose clerk deducted three per cent for keeping the accounts. Although each pilot received £480 for the year ending 30 June 1841, they demanded that the clerk's commission be reduced to one and a half per cent. James Osborne, the clerk, protested to the harbour master and claimed he 'had to run after the pilots from one pothouse to another to call them to attend ships about sailing in their respective turns'.

Captain Lewis was sympathetic, and obviously envious of the pilots' improved fortunes when he recorded:

> I must soon look up to these people for they are now buy-

ing land around me to establish their residences and country villas and will soon be disgustingly rich.

Improvement in their finances did not improve the pilots' behaviour. In August 1841 the *Port Phillip Patriot* newspaper complained that the pilot in the *Strathfieldsaye* prevented a reporter from boarding and had threatened to 'sink his boat if he did not lay off', although a reporter from the rival *Herald* had been allowed on board. Replying to the complaint the pilot claimed that reporters on board prior to anchoring caused confusion, which could result in the pilot being left to work the ship without an officer being present to enforce discipline. In the *Strathfieldsaye* the crew were in a state of mutiny and passengers were working the sails!

Apparently the argument had developed from a previous encounter between pilot and reporter when they jointly hired a boat to take them to ships anchored in Hobson's Bay. An argument ensued and on their return the reporter had abused the pilot and threatened to publish adverse reports about him in the newspaper. The pilot won the second round when reporters were prevented from boarding ships before they anchored, the pilots being enjoined to show no favour.

Although the *Ranger* was sent to Port Phillip as a cruising pilot ship she was not exclusively used in this capacity. Much of her time was occupied laying buoys and tending seamarks within Port Phillip. On 8 August 1841 a pilot arriving at Williamstown reported to Captain Lewis that the *Ranger* had been blown ashore on Swan Island in a gale. There were no pilots at Shortland's Bluff at the time and Captain Tobin was first on turn at Williamstown. Tobin was located 'drinking at the tap of a public house' and was ordered by the harbour master to proceed to Shortland's Bluff in the paddle steamer *Sea Horse* to endeavour to refloat the *Ranger*. Disregarding this instruction Tobin took another ship to sea and as a result Captain Lewis demanded that Tobin's name be 'erased from the list of pilots for his port'. Tobin escaped with a caution.

Meanwhile the *Ranger's* situation was precarious. High and

dry on a sandy beach, with rudder and keelson damaged, buoys intended for the West Channel were removed by the pilot's boatmen. Captain Lewis chartered the schooner *Boujah Maiden* at £4 a day, to run anchors which eventually refloated the stranded *Ranger*. *Boujah Maiden* was used to lay the buoys while the cutter was repaired.

Less than a month later Tobin and Lewis again crossed swords after the pilot anchored the brig *Deva* in Hobson's Bay without first heaving to for the harbour master and a health officer to board. Tobin, called to account for his action, blamed the squally weather and claimed that Captain Sproat of the *Deva*, accepted responsibility for the breach of port regulations. This time Tobin was suspended from duty for a month.

Apparently the port could not afford to lose the services of a pilot. Captain Sutton had resumed seafaring and was in command of the schooner *Devonshire* trading between Melbourne and Geelong. Lewis wrote to Sutton on 17 September 1841 instructing him to resume piloting.

On 4 October the shortage of pilots was further aggravated when McClelland was 'taken seriously ill of rheumatic gout and obliged to be sent in a cart across the country to Geelong for medical aid leaving but little hopes for his recovery'. This pessimistic view of McClelland's condition was apparently unwarranted because, on 17 December, his licence was withdrawn due to his carelessness in allowing the *Mary Nixon* to ground off Gellibrand's Point. And when the lighthouse was established on Shortland's Bluff, McClelland's name appears again as the original lighthouse keeper.

Early in 1842, four more pilots – Ruffle, Jacks, Thompson and Henry – were licensed; thus nine were operating. Captain McPherson was then domiciled at Williamstown and had adopted the title 'Master Pilot'. As such he did not consider himself liable to a share in the operating expenses at Shortland's Bluff though the other pilots thought otherwise and sought Superintendent La Trobe's opinion. McPherson was ordered to bear his full proportion of general expenses regardless of his title.

There is little information on the number of pilots operating between 1842 and 1851, except that several were working on a temporary basis. On many occasions no pilots were available at all and the masters of approaching ships stood on towards the lighthouse hoping their signals for a pilot would be answered. Several came to grief in the unexpected strength of the tide. One of these was the barque *Victory,* one hundred and six days out from Glasgow with sixty passengers. It grounded in fourteen feet of water between Point Lonsdale and Shortland's Bluff on 15 August 1850. This has since been known as Victory Shoal.

On 21 March 1852 Pilot Lawler was a little late leaving Shortland's Bluff in his whaleboat to board the Geelong bound four hundred and thirty-four ton emigrant ship *Isabella Watson.* The pilot was endeavouring to attract the captain's attention when the *Isabella Watson* struck what was later named Corsair Rock, and was swept on to the nearby Nepean Reef. Pilots Tobin, Toby, Taylor and Caught manned another whaleboat and went to the rescue. Tobin swam to the wreck with a life-line and was instrumental in saving the sixty passengers and all except nine of the crew. These men were drowned when a boat from the wreck capsized in the turbulent seas. Survivors were landed on Point Nepean whence the paddle steamer *Aphrasia* was sent to take them to Geelong.

Pilot Tobin's bravery was recognised when shipmasters in the port of Melbourne presented him with an inscribed 'chronometer watch' on behalf of the survivors from the *Isabella Watson.* Six months later he was in charge of the *Otillia* when she was struck by lightning in West Channel. Luggage and a pump at the foot of the mainmast were damaged and three of the crew injured by an explosion of gunpowder.

Soon afterwards the veteran pilot retired to unsuccessfully tender for the mail contract between Geelong and Shortland's Bluff. He then opened the first butcher's shop in the newly named township of Queenscliff. Tobin later moved to Sydney where he died on 11 September 1874.

On 15 September 1852 the schooner *Portland* and the ss *Conside* were both wrecked on Nepean Reef. The schooner grounded first and the steamer's captain, who was exempt from pilotage, steered for the stranded schooner's lights under the impression she had entered safely. Some unruly passengers were drowned when they took charge of one of the *Conside's* life-boats. There had also been some looting by the time the pilot's boats arrived to rescue the remainder of the passengers and crew at daylight.

Another ship which came to grief through entering without a pilot was the six hundred ton *St George,* arriving from London on 28 December 1852. After being swept on to Corsair Rock by the strong ebb tide she was beached and abandoned in Swan Bay.

CHAPTER 3

The Cruising Schooners

During the first session of the Legislative Assembly of Victoria in 1851 a bill was introduced 'to provide for the rate of pilotage in the several harbours in the colony of Victoria'. When the bill was sent to England for Royal Assent the Privy Council advised Lieutenant Governor La Trobe that:

> . . . because the Act affects not only local interests but the trade of other countries with the colonies, unless some amendment was introduced, or further explanation offered, the Queen would be advised to withhold Royal Assent.

The reason for the objection was a proposed 'tonnage duty', to be levied on all vessels whether a pilot was employed or not. Privy Councillors considered that pilots on salaries, independent of the services they rendered, would lack 'principal inducement to exertion'. Meanwhile the Port Phillip pilots had been placed on a salary basis by the new government of Victoria.

Although the *Ranger* was used as a cruising pilot ship no endeavour was made to establish a continuous cruising pilot station until 1852, when the government bought the ninety-nine ton brigantine *Boomerang* for the purpose. When this vessel had arrived at Melbourne from Glasgow, newspapers reported roomy accommodation far beyond what could be expected in so small a vessel.

After being refitted for her new role *Boomerang* sailed from Melbourne on 18 December 1852 under Captain Caught's command and with Lieutenant Governor La Trobe as a passenger, to inspect the Pilot Station at Shortland's Bluff and the Quarantine Station at Portsea.

James Gardiner Caught joined the pilot service under the salary scheme at the age of twenty-three. His name is first on the list of pilots licensed by the Pilot Board in 1854. Early in 1858 he left Queenscliff in a whaleboat to join the cruising pilot ship outside the Heads. Bad weather forced the whaleboat to put back. It capsized in doing so and one of the boatmen was drowned. Two others brought the unconscious Captain Caught ashore and revived him with artificial respiration. These two men, Fanning and McGrath, were later awarded Royal Humane Society medals for their efforts. Captain Caught was also aboard the *Rip* when she was dismasted in 1873. He resigned from the Pilot Service in 1880, died the following year and is buried in Point Lonsdale cemetery.

The continuous cruising pilot station was inaugurated by the *Boomerang* on 29 January 1853, when Pilot Ashby was put on board the five hundred and seven ton *Dolphin* arriving from New York. It has since been maintained.

When the Privy Council's objection to the compulsory pilotage rate reached Melbourne the proposed 'tonnage rate' was amended to include wharfage, water police charges and light dues. Justifying this action the lieutenant governor wrote to the colonial secretary:

> . . . owing to the high prices prevailing in the colony since the discovery of gold, the size of Port Phillip Bay, and the great distance for which pilots are required to take charge of vessels, pilot dues sufficient to repay pilots for their services are too heavy an impost on shipping and might prove injurious to the commercial interests of the port.
>
> . . . A rate sufficient to form the whole remuneration of the present staff of pilots would require the existing charges to be doubled.
>
> . . . The navigation of that portion of the harbour adjacent to the heads is difficult and intricate. The Council was desirous of compelling all save coasting vessels and those

Eastern light at the foot of Arthur's Seat

The Black Lighthouse with the Lookout Tower, now raised on a brick pedestal

The pilot steamer Akuna, formerly the German steam yacht Komet

The 40-foot pilot launch Hawk II

acquainted with the port to take pilots, viewing every shipwreck as dangerous to the character of the port.

Replying to the amended proposals the colonial secretary, the Duke of Newcastle, advised that, although there were some objections of a general nature Royal Assent for this Act would not be withheld.

All vessels calling at Victorian ports are still levied a 'tonnage rate' in addition to pilotage charges.

Shipping continued to converge upon Port Phillip from all corners of the globe. In October 1853 there was reputed to have been four hundred and sixty-five vessels anchored in Hobson's Bay. Twenty-six ships arrived off the Heads on 21 April 1853. Four days later there was a disastrous indication that the cruising pilot station would not prevent shipwrecks.

Early that morning the four hundred and fifty ton *Sacramento,* carrying three hundred emigrants as well as £60,000 in specie, drifted ashore near Point Lonsdale while hove to awaiting daylight. It was the signalman at Point Lonsdale who reported the stranding. Within two hours a whaleboat in charge of Pilot Mansfield was on the scene. After passengers and crew had been ferried ashore Mansfield returned to the stranded ship, which was lying well up the reef on her port side. With three boatmen he transferred ten boxes of specie to the schooner *Empire* for delivery to Melbourne.

Mansfield, who had been a pilot for only six weeks, resigned when he received a salvage award of £5000. James McIntyre, one of the boatmen, invested his £1500 award in the purchase of the brig *Montezuma* and subsequently pooled his resources with former pilot Captain James Deane, in what eventually became the Melbourne Steamship Company.

When, on 23 June 1853, the eight hundred and thirty-nine ton *Earl of Charlemont* was wrecked near Barwon Heads, in similar circumstances to the *Sacramento,* Pilot Taylor led a rescue party overland, swam the Barwon River and organised the rescue of passengers and crew. Eight bullock drays were sent from Geelong to convey the passengers and their belongings to their destination. Captain Taylor was another pilot who joined the service when the government took over its operation in 1851. He was killed in a buggy accident on the Geelong to Queenscliff Road in 1858; and is also buried at Point Lonsdale.

Nor were outward bound ships safe from disaster. On 30 May 1853 the eight hundred ton *Sea,* bound for Chile in ballast, anchored off Shortland's Bluff because Pilot Taylor did not consider it was safe for the *Boomerang* to take him off in the gale conditions. The captain of the *Sea* decided to sail without a pilot when the ebb tide made at 4 pm. Two hours later the *Sea* was ashore on Nepean Reef.

Captain Mackay and sixteen of his crew thought they had a better chance of survival if they stayed with the wreck. Seven others left in a boat and were rescued. When the *Boomerang* approached the wreck at daylight the seventeen men were marooned on the foredeck. Two pilot boats were swamped attempting a rescue and the *Boomerang's* steward sacrificed his life while endeavouring to swim ashore with a lifeline. None of the seventeen survived.

On 1 April 1854 another wreck was caused by the captain refusing to accept the pilot's authority. This was the one hundred and eighty ton *Columbine* on Ocean Grove beach. Five lives were lost, including Pilot Davis's.

It was these and other disasters that brought another change in the pilotage system. Six pilots were always to be available at Shortland's Bluff; exclusively to take ships in and out of the Heads. Only mail steamers and men-of-war could be piloted past Shortland's Bluff by an 'outside' pilot if there were no 'inside' pilots available. If the captain was not prepared to navigate South or West Channel without a pilot his ship was anchored until an inside pilot was available.

Pilotage rates were fixed at 3d per ton with a minimum of £4 and a maximum of £15 and had to be paid whether a pilot was employed or not. Another 1½d a ton was charged if a pilot was used, inwards or outwards, beyond Shortland's Bluff; the maximum then being £20. Vessels under tow, or returning to port within twenty-four hours of departure, were charged half rates. Fishing vessels and coasters were the only ships exempt.

There was a master pilot at Williamstown, a head pilot at Shortland's Bluff and a channel master in charge of the schooner *Apollo* responsible for buoys, beacons and seamarks.

Dues were paid at the Customs House before a ship cleared, and whether the 3d a ton compulsory rate, or the 4½d a ton 'all the way' rate was paid, it was endorsed on the clearance so the pilots could act accordingly.

Captain Ferguson, who had taken over from Captain Lewis as chief harbour master, was in control. His 'Instructions to Pilots' concluded:

> The respectability of the service will best be kept up by the pilots ever maintaining in their intercourse with each other, and the public, a courteous demeanour and habits of strict temperance.

Not long after the *Boomerang* was commissioned the one hundred and forty-six ton *Pacifico*, built in France in 1848, was acquired as a second cruising pilot ship. Experience with both ships proved that square riggers were not suitable and fore and aft rigged cruising pilot ships were sought.

At Sydney the former Boston, USA, pilot schooner *Anonyma*, which had proved her sailing qualities against the local cutter *Midge*, was bought by the Victorian government and arrived at Port Phillip to be used as a cruising pilot ship on 29 April 1853. Another suitable vessel, the cutter yacht *Corsair*, eighty-four tons, formerly of the Royal Yacht Squadron, arrived on 6 June, seventy-four days out from England. Impressed by her sailing qualities the pilots recommended the *Corsair* to Captain Ferguson as a cruising pilot ship. The Tasmanian built schooner *Prosperine* was also purchased at Hobart and suitably modified.

In August 1855 the *Boomerang* was sold to Theodore Sumner and went to Hobart. Captain Devlin bought the *Pacifico* the following year and sent her to Hong Kong. *Ranger* had been sold to Thomas Rimington in 1854.

With the rush of shipping attracted by the gold rush, more and more pilots were employed by the government; the numbers rising to as many as fifty. All had to be housed at Queenscliff. Pilot Row, a group of weatherboard houses at the southern end of Gellibrand Street, were the first houses built in the newly surveyed township.

Government records disclose that pilots' quarters cost £117 in 1852, £14,167 in 1853, £1210 in 1854, and £1525 in 1855. These costs, in addition to pilots salaries, and purchase and maintenance of the pilot vessels, caused the pilotage dues to soar to 1s 6d a ton for sailing ships and 10d a ton for steamships. In 1854 it was mutually agreed that the financial operation of the pilot service should be returned to the pilots.

Power to appoint pilots, frame regulations and dismiss anyone considered unworthy, was vested in a new Pilot Board of five, representing the government, the pilots, shipowners, Melbourne and Geelong merchants, with Captain Ferguson as president. Fees continued to be collected by the Customs Department but were now paid into a bank account operated by the Pilot Board. After expenses of operating the service had been paid the balance was divided among the pilots.

Some of the government employed pilots refused to accept the scheme and resigned. Twenty-six new pilots were licensed to replace them, bringing the total complement to forty-five. These men were required to purchase the *Anonyma*, *Corsair* and *Prosperine* from the government at auction, each contributing approximately £85 towards the cost.

Within a year the rates had been reduced to 5½d a ton for sailing ships, 3½d for steamships. A tonnage rate continued to be levied by the government in addition to pilotage fees.

Early in 1855 the Pilot Board introduced a new set of rules and financial arrangements. Seven companies of sea pilots were formed. Harbour pilots operated a separate service at an additional cost of 1d per ton for vessels proceeding to Melbourne, Footscray or Yarraville wharves. Harbour pilots operating their own boats, and not sharing the sea pilots establishment, operated until 1957.

During the first nine months' operations the Pilot Board handled £43,000; paid £15,295 in expenses, including cost of pilot ships and plant. Each pilot received an average of £720. Working expenses for the first six months of 1855 were £5681.

Six pilots continued to be stationed at Queenscliff to ensure that arriving ships would always have a pilot available, but the system of changing pilots at Shortland's Bluff was discontinued.

In 1854 the two thousand five hundred and fifteen ton *James Baines* was the largest full rigged ship afloat. Captain McDonald was in command on her maiden voyage. After embarking a pilot off Port Phillip Heads, sixty-five days after clearing Liverpool, he refused to heave to off Shortland's Bluff to land the Geelong and Western Districts mail and was fined £100 on arrival at Melbourne for this breach of port regulations.

When the *James Baines* arrived on her second voyage Pilot James Deane was embarked at 1 am on 2 October 1855, twenty miles south-west of Shortland's Bluff. Despite a strong northerly the lighthouse was abeam by 9 am. Captain McDonald again refused to heave to to land the mails and did so only after the pilot had threatened to anchor the ship. After the mail had been landed the *James Baines* proceeded via the South Channel, tacked against an ever increasing northerly and arrived in Hobson's Bay in the remarkable time of eight and a half hours.

Next morning the *Argus* newspaper paid Pilot Deane this compliment:

> We think it is due to Mr Pilot Deane to remark upon the skill and perseverance he used in beating the *James Baines* into Hobson's Bay against a head wind. We cannot at all times compliment the pilots — indeed there have been but two or three occasions when we have ventured to do so . . .

Soon after Captain Deane joined the pilot service in 1853 he acquired a financial interest in tugboat operations. Pilots could benefit tug operators offering towage assistance to sailing ships arriving outside the Heads by persuading captains to accept a tow to their destination. In 1866 James Deane resigned from the Pilot Service to attend to his increasing interest in steam tugs; an interest which developed into the Melbourne Towage and Lighterage Company and eventually into the Melbourne Steamship Company.

Deane's dual operations resulted in the Pilot Board introducing a regulation – which is still in force – prohibiting Port Phillip Pilots from holding 'directly or indirectly any share or interest in any steam tug, or any steam vessel used regularly or occasionally for towing vessels within the port for which he holds a licence'.

Captain Ferguson insisted on three pilot ships being maintained at the Heads. Two of these were always to cruise outside; one twenty miles off, the other five to ten miles off. The former to supply pilots to any inward bound ships encountered, the latter to take pilots off outward bound ships and to supply pilots to any that missed the ship on the outer station. The third pilot ship always to be manned and ready inside the Heads. The three ships changed stations alternately.

Pilots had the right to take to sea any vessel they piloted inwards. If not available the pilot first on turn for his particular company took his place. Each company operated a separate fund and there was healthy competition.

Sometimes the competition was too healthy. In an endeavour to obtain customers the outside cruising ship might go as far as Cape Otway, from which direction most of the big tonnage arrived, leaving the eastern approaches uncovered. Captain Ferguson introduced cruising limits to counter this tendency.

Before long the seven companies were reduced to three, each owning one of the pilot ships: No 1 the *Corsair,* No 2 the *Proserpine,* and No 3 the *Anonyma.* The last pilot to board the ship took charge and each had an apprentice pilot as first mate. With the decline in shipping there were fewer vacancies in the service and some of the apprentices had to wait several years for promotion.

It was an apprentice pilot who allowed the *Anonyma* to drift into the breakers west of Point Lonsdale on the morning of 2 October 1859. No lives were lost but the ship was wrecked. Needless to say that apprentice did not wait any longer for a vacancy. At the time of her loss the *Anonyma* had proved so successful that No 2 Company had ordered a similar but slightly larger vessel as a replacement for the *Proserpine.* This vessel, to be named *Rip,* was ninety-four feet in length, twenty-three feet beam, with a main-mast eighty feet long and a main boom forty-six feet; dimensions almost identical to the schooner-yacht *America,* which was designed on the lines of New York pilot schooners. In 1853 *America* outsailed British yachts and gave her name to the famous America's Cup.

The *Rip* was built at St John, New Brunswick, and took one hundred and twenty-two days to sail to Port Phillip, where she arrived on 2 May 1860. Workmanship was deemed to be 'rather rough' but the model perfect. Although she performed well on her delivery voyage the spars were reduced in size after arrival. Total cost, including alterations was £4500. Trials off Williamstown on 2 June were marred by a flat calm and the twenty-three guests on board for the occasion were disappointed when she had to be towed back to her berth.

After the *Anonyma* was lost the cutter *Swan* was chartered until the *Rip* arrived. This vessel was built in France in 1815, first arrived at Hobart in 1837 and had been trading across Bass Strait. She brought George Coppin's theatrical company to Melbourne in 1846; was still afloat in 1906 when, ketch rigged, she foundered after a collision off Cape Schanck.

Shipping had fallen off by 1861 and the pilots decided that two ships were sufficient, with the *Proserpine* in reserve. One maintained a cruising station fifteen miles from the Heads while the other held an inside station off Shortland's Bluff. Two companies of eleven pilots each were able to cope with the reduced shipping. But the three ships were maintained in first class order.

Shipwrights at Queenscliff worked for both companies. Ownership of the *Proserpine* was transferred to 'Charles Ferguson and others'. When the *Rip* or the *Corsair* were under overhaul *Proserpine* was chartered by the company using her at a cost of £5 a day.

When the clipper *Empress of the Seas* sailed from Melbourne on 19 December 1861 with cargo including gold to the value of £80,000, some of the crew were reported to have threatened to destroy the ship on the homeward voyage. Captain Bragg decided to anchor off Shortland's Bluff while he went to Geelong to obtain ten seamen to replace the troublesome element. It was late at night when he returned to Queenscliff with the replacements, and all were accommodated for the night at an hotel. At about four o'clock in the morning Captain Bragg was awakened with the news that his ship was in flames. Fire had broken out in the forehold and spread aft. Hoses were coupled to fight the flames but no trace of the pump handles could be found. Apparently the mutinous element in the crew had thrown them overboard. The only alternative was to try to fight the fire with buckets but the ship was soon competely enveloped in flames.

In the absence of Captain Bragg another shipmaster, Captain Brown, a passenger in the *Empress of the Seas,* took charge and supervised getting the passengers ashore in the ship's boats. When he tried to slip the anchor cable to beach the ship he was prevented from doing so by the blazing foremast, which had crashed to the deck.

Eventually the anchor cable was slipped and as the burning ship drifted seawards on the ebb tide, the gold, stowed right aft in the lazaret, was brought on deck. Captain James Deane, in charge of the *Rip,* came to the rescue. Twenty-two boxes of gold and some other valuables rescued from the blazing ship were taken aboard from one of the *Empress of the Seas* boats in the vicinity of Corsair Rock.

Empress of the Seas was beached on the Nepean shore inside the Heads but was eventually towed away to be broken up.

CHAPTER 4

Triumph and Disaster

When Cape Otway lighthouse was connected to the electric telegraph system arrivals from the west could be anticipated. Each vessel maintained the outside station for eleven days, obtaining inward pilots from outward bound ships whenever possible; and from Queenscliff at other times.

There was still some competition. After all available pilots had been shipped the outside pilot ship had to haul down her commission flag, a signal that the mate of the inside ship must ever be on the alert for. Then he would round up pilots to take over the outside station until the other ship was ready to resume.

An eleven day station would commence with the ship taking over hoisting her flag at noon exactly. There might be six pilots on board, the last on turn being in charge. In favourable weather she would beat to windward and then drift back with sails ashake. In heavy weather she would lie hove to under storm canvas.

At night a bright white light at the masthead, and a flashing light every fifteen minutes from the deck, identified her to approaching ships. No sidelights were carried because a pilot ship was required to keep clear of other ships. Should a steamer's masthead light, or a sailing ship's sidelights, be sighted, the pilot ship would sail towards her, keeping a lookout for the blue flares: the signal made by a vessel requiring a pilot. When this was identified the ship's name and draught would be ascertained while she hove to for the pilot ship to round her stern and come up into the lee to send a boat.

If more vessels were expected than could be supplied with

pilots from outward ships the numbers would be made up by pilots coming from Melbourne by tug or steamer. In addition the six pilots domiciled at Queenscliff were always available for unexpected arrivals.

While on the inside station the pilot ship would undergo overhaul, with the first mate in charge to watch his company's interests. There was a lookout tower on Shortland's Bluff with a powerful telescope to assist. Any ships entering the Heads with a signal flying for a pilot were customers for the inside ship. Maybe on the fourth day the mate of the inside ship would inform the pilot first on turn at Queenscliff that there were only two pilots on the outside ship and that four ships had been reported passing Cape Otway, running before a fresh southwester. This should indicate that the inside ship should be ready to sail when the ebb tide made at midnight. Perhaps three Queenscliff pilots would be on board. If the southwester approached gale force, storm canvas would be set, for beating seawards with the tide but against the wind.

Approaching Point Lonsdale the seas steepen, heralding the turbulence of the Rip. 'Haul out the bowline! Put the helm down!' the pilot in charge orders. The ship spins round on to the other tack, hove to with the headsails to windward.

'Lay aft!' is the next order. The companionway is battened down and all hands grab a firm hand-hold. Pitching and rolling violently the ten knob ebb sweeps the little ship southeastwards, towards the most turbulent water. 'Look out men!' follows, as the pilot in charge prepares to resume sailing.

'Let draw!' he yells. The jib sheets are run to leeward, the sails fill on the port tack and the little ship heads westwards towards the calmer water beyond the line of the main leading lights, close-hauled against the southwest wind.

By daylight she might be eight miles offshore, close to the other pilot ship. Approaching is a steamer with the signal flying for a pilot. There is another steamer in the offing. When these two ships have been boarded the mate of the inside ship's alertness is rewarded. Down comes the outside pilot ship's flag. There is another ship approaching from the south-west. Up goes the flag on the relieving ship.

This is the time for decision by the mate of the outside ship. With no pilots on board he has to decide whether to go into Shortland's Bluff for more, or wait outside to replenish his supply from outward bound ships. He decides to stay, and watches while the rival ship steals a customer, worth probably £40 to his company. That is all they get. A plume of smoke near Point Lonsdale develops into a tug with the flag of the outside company's ship flying.

Those aboard the inside ship strain their eyes to see how many pilots are aboard the tug, but the pilots coming out know the score and keep out of sight. Heads are counted as they transfer by boat from the tug to the outside ship. Up goes that ship's flag. Down must come the flag of the inside ship, which returns to the anchorage off Shortland's Bluff.

On 1 January 1868 this kind of competition ceased when the two companies decided to pool their earnings. But lack of competition revealed other weaknesses and within six months the pilots had reverted to the eleven-day station system, to strengthen the work once more.

Vessels trading intercolonially and fishing vessels were exempt from pilotage fees if their captains could prove their local knowledge by examination at Melbourne. Ships so exempt still have to fly a large white flag to identify them to the pilot ship.

The fee for the exemption examination, payable to the Pilot Board, was £20. These fees could not be claimed by individual pilots, and there was no provision for them to be paid into the Treasury account. By the end of 1857 such fees amounted to £2761 and this established what became known as the Exemption Fund.

When the pilots ceased to be government employees they forfeited superannuation rights. It was decided that the Exemption Fund would be applied as compensation for this forfeiture. However, there had been laxity in issuing receipts for monies paid into the Exemption Fund. By the end of 1860 there was more than a suspicion that the secretary of the Pilot Board had embezzled at least £6000. Although the pilots reported their suspicions, it was 1874 before a Select Committee was appointed to investigate. The report of this Select Committee upheld the pilots' claims and recommended that a sum of £4344 be paid into the Exemption Fund by the government as compensation.

By the end of 1870 the fund, which had become known as the Pilot Board Fund, totalled £13,200 and was accruing £1350 a year from interest and other sources. Twenty-five pilots held an interest and one widow's pension and five children's allowances were being paid.

From 15 June 1871 the Pilot Board Fund became the Port Phillip Pilots Sickness and Superannuation Fund, financed by three and a half per cent of the total pilotage fees before any disbursements were made. It still exists in this form and is administered by the State Government of Victoria. Contributions are now six per cent of the total pilotage fees. Original provision was for pilots to receive a minimum pension of £30 a year at the age of sixty after ten years service, and a maximum of £130 a year after thirty-five years service. Widows to receive a minimum of £20 and a maximum of £70. Pilots who left the service under the age of sixty were allowed a retiring allowance of £22 10s for each completed year of service as settlement of their claim on the fund. A similar system operates today, upgraded to modern standards.

Shipping continued to decline and there were no vacancies in the service for many years. Then, on 26 November 1871, Pilot Hansen was aboard the schooner *Belle* bound for Hobson's Bay when the vessel was struck by lightning. Captain Hill, who had been standing close by, was horrified to see the pilot dead at his feet, the imprint of the pilot's hand burnt into the schooner's taffrail. John McKenzie, mate of the *Corsair* for fourteen years, filled the vacancy caused by Captain Hansen's death; and lost his life in even more tragic circumstances within two years.

On 15 July 1873 the *Rip*, in charge of veteran pilot Best, with five other pilots on board, slipped her moorings off Shortland's Bluff at about 9 am. With the tide ebbing strongly against a south-west gale, seas were breaking across the entrance. Captain Best elected to remain inside until the tide eased.

Approaching slack water at 11 am there was some improvement. Under storm canvas the schooner headed towards the Rip, heaving to with the storm jib to windward and drifting out through the breaking seas on the last of the ebb. Clear of Point Nepean, when the order 'Let draw' was given, the sails filled on the port tack as three huge tiers of breaking seas swept in, carrying away the mainmast as the storm trysail filled with water.

Four men were swept overboard as the schooner was pushed on her beam ends to starboard: Pilot McKenzie, Classon the cook, Wells the steward and James Marr an AB. The *Rip* slowly righted, revealing James Marr clinging to the rigging of the trailing mainmast wreckage.

All, including Marr, realised that if the *Rip* was to survive, the trailing wreckage would have to be cut adrift. Marr sacrificed his slim hope of survival when he yelled to his shipmates to cut him adrift. The flood tide was making and Corsair Rock close alee. Free of the trailing wreckage, with fore storm trysail and storm jib hard on the wind, the pinnacle of Corsair Rock was weathered by a mere fifty feet and the *Rip* regained the safety of Shortland's Bluff anchorage.

Four lives had been lost in the worst disaster ever to befall the Port Phillip Pilot Service. Internally the *Rip* had sustained heavy damage. Bulkheads were flattened and wrecked internal fittings spread over the ballast. Repairs cost £1600.

An inquiry by the Pilot Board decided that circumstances over which the pilots had no control caused the disaster, and praised all on board for their conduct. Publicity given to James Marr's sacrifice was the direct cause of George Coppin, MLC, urging the formation of the Victorian Humane Society which came into being on 28 September 1874, exactly one hundred years after the Royal Humane Society was founded in England. Eight years later the Victorian Society was merged into the Royal Humane Society of Australasia.

Only ten months after the *Rip* disaster the *Corsair* was lost on Nepean Reef in similar circumstances but without loss of life. In May 1874 a series of disastrous gales wrought havoc upon shipping in the Southern Ocean. *The British Admiral* was wrecked on King Island, and four large ships – *Loch Ard, John Kerr, Cambridgeshire* and *Dallam Tower* – arrived off Port

Phillip Heads in a dismasted condition. These and other vessels delayed by the gales arrived simultaneously on 24 May.

There were not sufficient pilots on the outside station and the *Corsair* had to leave the cruising station to go inside for more pilots while the ships waited. Beating seawards again against the first of the flood tide she missed stays off Corsair Rock and was carried by the tide on to Nepean Reef. James Deane's tug *War Hawk* left the dismasted *Loch Ard* to try to pass a line to the *Corsair,* but was unable to get close to the stranded cutter, which was bilged and waterlogged.

Pilots Kennedy and Robertson, with one AB, launched one of the workboats and spent an uncomfortable night at sea before landing safely. Another boat manned by five pilots and the remaining six crew members of the *Corsair* managed to reach Queenscliff safely the same night. The government steamer *Pharos* took over the cruising pilot station until the ketch *Mimosa* was chartered by the pilots to take the *Corsair's* place.

When the *Corsair* was wrecked Captain Deane was about to leave for England in a quest for more steam tugs. He undertook the task of seeking a replacement for the pilot cutter. At Glasgow, Deane located the twenty year old, seventy ton schooner yacht *Mavis* – built in 1852 – for sale and in very good condition. Deane bought the *Mavis* for No 1 Company and enlisted a crew to sail her to Melbourne.

Soon after departure the crew demanded that Captain Elliott, the delivery master, put into Falmouth because the schooner's lofty masts and lengthy spars were unsuitable for the proposed voyage. After these had been reduced to the crew's satisfaction the *Mavis* cleared Falmouth on 8 September 1875, performed satisfactorily in Biscay gales, rounded the Cape of Good Hope on 2 November, was forced to heave to for about a week in the Southern Ocean by bad weather, but safely entered Port Phillip on 16 December.

Reporting on the schooner's behaviour Captain Elliott said her best day's run was two hundred and fifty miles. When running, hove to, or on the wind, she shipped no water.

Cost of the *Mavis,* including alterations and delivery, was close to £5000. She proved to be the equal of the *Rip* in most respects.

One of her best achievements was a record six hour run from the cruising station to Williamstown and return; a distance of nearly seventy miles.

In July 1879 two large sailing ships, *Oberon* and *Merwanje Framjee,* arrived at the pilot station in a southwest gale which had forced the pilot ship to shelter. Both ships hove to late in the afternoon, but as conditions worsened towards nightfall the captain of the *Merwanje Framjee* clawed offshore and returned the following morning. *Oberon's* captain, unable to beat seawards, decided to run through the Heads without a pilot and was boarded in smoother water inside.

Inquiring into the complaints made by these two shipmasters the Pilot Board found that the pilot in charge was justified in sheltering. But it was this incident that led to eight local shipmasters applying to the government for permission to fit out a suitable vessel in which to cruise beyond the existing pilotage limits; and for licences as Port Phillip pilots.

Commenting on the proposal, Captain Charles Payne, president of the Pilot Board, said that another cruising pilot ship was unnecessary. He pointed out that during the previous five years twenty-six thousand vessels, aggregating ten million tons of shipping, had passed through the Heads with pilots and there had been only thirty-six casualties. Another pilot service, with no obligation to patrol the approaches, could select big ships and leave the less lucrative small tonnage for the regular pilots.

Several of the applicants, who were sponsored by the Chamber of Commerce, had been rejected when they applied for vacancies in the pilot service, mainly because of age. At that time the maximum age for pilots joining the service was thirty-five. The average age of the six pilots appointed during the preceding nineteen years was thirty-three. Twenty-six overseas shipmasters and twenty shipowners, or their Melbourne agents, supported the eight in their quest to fit out a suitable, vessel. But on 25 November the government turned down the proposal.

This was only one of the complaints from the Chamber of Commerce about the inadequacy of the pilot service. Two years previously a Bill sponsored by the Chamber and presented to

St Kilda from Hobson's Bay in the 1870s

Parliament proposing that the Pilot Board be replaced by a Marine Board similar to those operating in other states was not sufficiently supported. Both questions were raised again in 1886.

Captain Charles Payne was president of the Pilot Board from 1870 to 1880. Formerly a Royal Navy officer on the Australian station he became Inspector of Steamships at Melbourne in 1854, was secretary and Chief Inspector of the Victorian Steam Navigation Board before going to England to command HMS *Nelson* for the Victorian Navy. After Captain Ferguson's death in 1870 he became chief harbour master and president of the Pilot Board.

Captain Robert Fullarton took over this dual role when Captain Payne retired in 1880. This former Scot arrived in Melbourne in the fever-ridden ship *Glenbervie* in 1849, joined the harbour master's department in 1854, was in command of ss *Pharos* when appointed harbour master at Melbourne in 1869 and took over from Captain Payne as chief harbour master in 1880. He was also senior officer in the Victorian Naval Reserve with which he had been associated since the formation of the Williamstown Naval Brigade in 1859.

Captain William Nicholson, in the *Empire,* rendered valuable assistance to many vessels in trouble in and around Port Phillip. The *Empire's* normal function was to tend buoys and beacons and to keep the lighthouse keepers and the pilots stored and provisioned. Captain Nicholson also trained pilots on probation. When ss *Pharos* was built at Williamstown in 1867 to replace the *Empire* he was appointed harbour master at Geelong. He died in 1875.

In 1886 there were two companies of pilots operating, each comprising eleven pilots. No 1 Company owned the *Mavis,* No 2 the *Rip.* Each pilot's share was valued at £450 on which he received ten per cent interest per annum plus £2 a week for expenses and a further allowance for insurance, and it took the wreck of the four hundred and eighty-two ton French barque *Glaneuse* on 2 October 1886 again attract interest to the service.

This vessel arrived on the pilot station at 1 am in fine weather.

The captain claimed he made signals for a pilot but that the schooner on station sailed past and into port.

While the *Glaneuse* was hove to awaiting the schooner's return she drifted on to Lonsdale reef and was wrecked. The pilots claimed that if the French barque had been showing the proper signals a pilot would have been put on board on arrival.

This controversy led to the pilots demanding a public inquiry. On 19 November Captain W. Howard Smith presided over the Board of Inquiry which was established to consider a system of competition, the relative merits of steam and sailing pilot ships and whether or not the existing system of piloting was efficient. Captain Archibald Currie represented shipowners, assisted by Captain Underwood, a senior master in the Union SS Company's ss *Rotomahana.* Meetings were then being held in the Customs House at Melbourne.

Captain Currie was a fifty-five year old Scotsman who arrived in Hobson's Bay as an officer of the *Sir William Molesworth* in 1853. He left for the goldfields but soon returned to seafaring. After commanding his own ship at the age of twenty-two he established Archibald Currie and Company in 1872 and by 1885 had four large steamships in the Indian trade from Melbourne. Part of his fortune also came from salvage operations. When the *Netherby* was wrecked on King Island in 1886 he was one of the co-purchasers of the wreck and gave his name to Currie Harbour near the scene of the disaster. Other salvage successes included the *Hereford* on the Otway coast and the *Lubra* at Warrnambool.

The British India Line took over Archibald Currie and Company in 1913.

Evidence before the Board of Inquiry into the Pilot Service disclosed that, in 1879, the pilots averaged £444 income and by 1882, this had risen to over £1000. Captain Payne, recalled from retirement to give evidence, considered the system satisfactory. If a second cruising pilot ship was to be re-established he asked where it was to be stationed. With steamships now comprising a large proportion of the arrivals a pilot station fifteen miles from Point Lonsdale would be impracticable.

Several shipmasters also gave evidence on the efficiency of the existing pilotage arrangements, and another thirty-eight shipmasters sent letters to the same effect. Leading shipping and commercial companies, and another group of shipmasters, signed a petition requesting a third pilot ship.

Victoria's Attorney-General and the Commissoner for Trade and Customs said that exhaustive inquiries showed that no improvement could be possible immediately, but that a Marine Board Bill was again under consideration by Parliament which was likely to reform the pilot service. This Bill was now opposed by the Chamber of Commerce because it was not intended that the Chamber should appoint a majority of members to the proposed Marine Board.

It was pointed out that Pilot Board Regulations provided that: 'No person could be appointed pilot who had not been master of a vessel over two hundred and fifty tons for twelve months, or chief officer of a foreign going vessel'.

On 29 December the Board of Inquiry announced that they considered twenty-four pilots sufficient but recommended that a third pilot vessel be obtained. Disapproval was expressed at the practice of the pilot ship leaving the station to come inside every day. The third ship was necessary to ensure that the cruising station should always be maintained. One ship was to cruise five to ten miles offshore, the other within five miles. Minor changes were suggested in the appointment of pilots and it was recommended that a lighthouse be established on Split Point.

Soon after the Board delivered their findings, the revenue schooner *Hawk* was advertised for sale at Auckland. Built for the New Zealand government in 1881, the *Hawk* had pohutakawa frames, kauri planks, beams and decks, and was copper fastened. Length eighty-five feet, beam twenty feet, draught nine feet, eighty-four tons register. Pilot Press went to Auckland to inspect the *Hawk*, decided she would make a good pilot ship and on 24 June 1887 successfully offered £2300 for her as she lay.

Captain Press sailed from Auckland in the *Hawk* on 20 July and on arrival at Williamstown twenty-one days later reported: 'It howled for two days but the schooner behaved well, never wet her knighthead. They said she would not stay but I tried her under staysail and she came round like a pivot'. On the ninety-six mile leg of the voyage, Wilson's Promontory to Point Lonsdale, she averaged ten knots. As a cruising pilot ship the *Hawk* had one fault; she would not heave to and had to be kept under way at all times.

While the *Hawk* was being delivered another shipwreck focussed attention on the pilot service. On 23 July 1887 the one thousand and seventy-one ton French barque *Gange* arrived at the pilot station just as darkness fell at 7 pm. The pilot in charge of the *Mavis* sent several signals to the *Gange* which were ignored. The barque sailed on and grounded on Lonsdale Reef where she became a total wreck.

The three schooners, *Rip, Mavis* and *Hawk,* maintained a continuous cruising pilot station off Port Phillip Heads until steam inevitably superseded sail in the early 1900s. When the *Rip* was sold she was converted into a fishing vessel and went to Western Australia. Later she became a lighter at Albany until worms in her planking rendered her unseaworthy. After being towed to sea and set on fire the charred remains drifted ashore on Frenchman Island, King George Sound, where they lie buried in the sand dunes.

Nothing is known of the end of the *Mavis.* In 1908 she was last registered at Sydney where it is possible she reverted to her original role as a yacht. *Hawk* was the last sailing vessel used by the Port Phillip Pilots. In 1903 she was sold, converted into a ketch, and also became a fishing vessel. Painted sky blue and red, with new sails and a figurehead, she was conspicuous in Victorian waters until she went to South Australia to be converted into a lighter sixty-seven years after being launched.

When the Marine Board Act of 1887 became law the rights of the Port Phillip Pilots to operate the service independently and to maintain their own establishments at Queenscliff and Williamstown were preserved. The new Marine Board, besides being the licensing authority for the pilots, took over the functions of the Victorian Steam Navigation Board of issuing certificates for masters, mates and engineers; surveying ships and fishing boats;

and investigating casualties. In turn the Marine Board is responsible to the Public Works Department of the State government of Victoria.

The Marine Board Bill continued to be opposed by the Chamber of Commerce. Although approving in principle they still demanded the right to nominate sufficient members to create a majority on the Marine Board.

Royal Assent to the Bill was granted on 17 December 1887. The Marine Board of Victoria has since been composed of twelve members: three appointed by the Governor in Council; two elected from the Roll of Merchants and Traders; one from the Roll of Underwriters; two from the owners of ships registered in Victoria; one by licensed pilots; one by shipmasters; one by ships' engineers and one by seamen in locally registered ships.

CHAPTER 5

Fairways and Foul Winds

Captain Lonsdale, Melbourne's first resident magistrate, arrived at Port Phillip in HMS *Rattlesnake* in 1836. Hobson's Bay was named after Captain William Hobson in command of the *Rattlesnake*. When his officers made an extensive survey of Port Phillip one of the prominent geographical features named was Shortland's Bluff, the middle headland of the entrance. It was named after Frederick Shortland, master's mate, who assisted Lieutenant Symonds to survey that portion of the bay.

In March 1837 the *Rattlesnake* returned to Port Phillip with Sir Richard Bourke, governor of New South Wales. Captain Hobson anchored off what is now Portsea to allow Sir Richard's party to seek a site for a lighthouse, a customs house and a pilot station. Sir Richard selected the summit of one of the sandhills east of Point Nepean as suitable for the lighthouse, and Shortland's Bluff as the most convenient position for a customs house and pilot station. But when he arrived at the settlement which he named Melbourne local shipmasters suggested that the lighthouse should also be on Shortland's Bluff.

Captains Tobin and Sutton, the first licenced pilots, used whaleboats launched from the beach in the lee of Shortland's Bluff to board ships. By 1841 the four pilots operating employed five boatmen and lived in tents on the beach until a carpenter was engaged to build their first permanent residence.

Paddle steamers were operating within Port Phillip in 1851, commencing with a service between Melbourne and Geelong by the *Aphrasia* and the *Vesta,* the latter being subsequently bought by Captain G. W. Cole to develop land he owned at St

Leonards into a seaside resort. To sell the land Captain Cole promised to establish steamship communication between Melbourne and St Leonards.

When the *St George* was beached and abandoned in Swan Bay early in 1853, after striking Corsair Rock, Cole bought the hulk and towed it to St Leonards where it was used as a landing stage for the *Vesta.* Later the steamship service was extended to Queenscliff and Portsea, and the channel from St Leonards to Queenscliff was named Coles Channel. When the government took over the lease of the cattle station at Shortland's Bluff, the steamship service, and a road to Geelong, led to the township of Queenscliff also being established as a seaside resort. Soon afterwards the port health officer's base at Portsea was transferred to Queenscliff and the Doctor's Jetty, now the Pilot Jetty, was built.

Early in 1849, an irregularity in the Shortland's Bluff light was blamed for the wreck of the four hundred and sixty-two ton barque *Princess Royal* on Lonsdale Reef while awaiting daylight to enter. Apparently the light had been temporarily shown from a platform above the pilots' houses. The captain of the *Princess Royal* blamed the consequent obscuring of the light for the wreck of his ship. The following year the *Victory* grounded between Point Lonsdale and Shortland's Bluff. It was towed off by the paddle steamers *Aphrasia* and *Vesta.*

On 28 April 1853, only three days after Pilot Mansfield had figured prominently in the salvage of the specie from the wreck of the *Sacramento,* he was involved in another adventure after boarding the four hundred and fifty ton American ss *West Wind,* sighted from the pilot ship at daylight flying signals of distress. Fuel was exhausted and ninety-five fathoms of anchorless chains trailed from the hawse pipes.

In July of the previous year the *West Wind's* owner, Adam Smith, advertised in New York

> . . . the palatial screw steamer *West Wind* for Port Phillip, with superior accommodation for sixty passengers and one

hundred and fifty separate berths in the steerage each with a porthole.

After the ship had sailed in September, with all berths filled, inefficiency, drunkenness and engine failures caused repeated delays. Coal supplies ran out in the South Atlantic and the *West Wind* was sailed to Ascension Island where she remained for sixty-three days awaiting fuel. At Cape Town, on 10 February 1853, passengers complained to the United State Consul that the *West Wind* was unseaworthy. Surveyors demanded major repairs, the captain was dismissed and Smith took command himself, resuming the voyage after another month's delay.

Approaching Port Phillip on the afternoon of 27 April the coal bunkers were again empty and the disabled ship drifted shorewards. Both anchors were dropped but the cables parted. The one hundred and fifty-three passengers had been on board for two hundred and thirty-three days. Wooden fittings were burned to raise steam to complete the voyage to Hobson's Bay. On arrival two guns were attached to the anchor chains to serve as anchors.

This was the first American steamship to visit Melbourne, preceding the ill-fated *Monumental City* by a few days. She made a coastal voyage to Sydney and Adelaide, returned to Port Phillip on 14 August and collided with the *Anonyma* while embarking a pilot, causing extensive damage to the pilot schooner. Unable to pay for these repairs Smith endeavoured to sell the ship. There were no buyers so he raised £2000 on a lien, repayable in three months. Two more coastal voyages proved unprofitable and the *West Wind* was laid up at Melbourne the following February.

Two weeks later she was destroyed by fire and the burnt out wreck beached near Sandridge, where for many years the boilers became a prominent landmark for pilots. Subsequently Smith, the purser, and the bosun were found guilty of setting fire to the *West Wind* after a trial which also exposed some well-known Melbourne businessmen who were suspected of being involved.

During the winter of 1854 Pilot W. Nicholson boarded the Portuguese barque *Tremalga*, arriving from Hong Kong with four hundred Chinese emigrants, short of provisions and unseaworthy. With a northerly gale raging and night approaching, the pilot decided to heave to until morning.

There was no sign of the *Tremalga* at daylight and the *Prosperine* cruised as far as King Island without sighting her. Five weeks later Pilot Nicholson turned up in ss *Waratah* from Eden. After being blown offshore in the northerly gale, the *Tremalga* battled against head winds for three weeks before running into Twofold Bay for shelter. By then pilot, passengers and crew were on the brink of starvation and the Chinese passengers in a state of mutiny. Another three weeks elapsed before the barque completed her voyage to Melbourne.

Pilot Patterson boarded the fourteen hundred ton clipper ship *Prince of the Seas* when she arrived on her maiden voyage on 29 December 1854 in squally northerly weather. Entering on the flood tide about 6 pm she was overtaken by a southerly buster which drove her on to Corsair Rock. She had to be beached inside the Heads to prevent her from sinking, and the cargo discharged before she could be refloated. After several successful voyages the *Prince of the Seas* was also burnt out at Sandridge a few years later.

Pilot Singleton boarded the ship *Lawrence Frost* arriving from Liverpool at 10 am on 17 August 1856. Owing to a strong northerly it was decided to anchor off Portsea. Approaching the anchorage the pilot ordered 'Square the main yard!' The carpenter misunderstood the order and let the anchor go. Unmanageable in the northerly and the strong flood tide the ship was aground within fifteen minutes. Badly damaged and flooded, it was a week before she was refloated and towed to Melbourne, where two thousand seven hundred tons of damaged cargo was unloaded.

There were no drydocks or slipways where repairs could be affected at Melbourne and the ship was ballasted with sand for the voyage to Sydney. This voyage was never completed as *Lawrence Frost* had to put into Twofold Bay in a leaking condition. She was beached and never refloated.

During the winter of 1859 the *Prosperine* grounded in the West Channel while bound for Williamstown. SS *Royal Shepherd* went to her assistance but had to return to Queenscliff to arrange for a tug, and brought two whaleboats back to the schooner, which was refloated by the tug *Black Eagle*.

Isambard Kingdom Brunel designed the famous ss *Great Britain* for the Great Western Steamship Company's North Atlantic run – the first ocean-going ship to be built of iron. On her third voyage from Liverpool to New York in May 1846, the *Great Britain* ran aground off the north-east coast of Ireland and it was fifteen months before she was refloated.

In Liverpool the cost of repairs was assessed at £22,000. The ship, which had cost £140,000 to build, was only insured for £17,000 and the company was forced into liquidation. Gibbs Bright and Company bought the *Great Britain* in 1850 for £18,000. Repaired and re-rigged as an auxiliary steamship she was the first regular sailing from Liverpool for the Australian Royal Mail Steam Navigation Company, departing on 21 August 1852 with six hundred and thirty passengers, a crew of one hundred and thirty-eight and two thousand one hundred and fifty tons of coal.

When she arrived off Port Phillip Heads on her third voyage in 1854, sixty-five days out from Liverpool, the port health officer discovered an outbreak of smallpox on board and sent her to the quarantine anchorage off Portsea for three weeks.

Portsea was not connected to Melbourne by electric telegraph and the *Great Britain's* belated arrival in Hobson's Bay was not anticipated. Melbourne residents knew of Britain being at war with Russia in the Crimea and there were rumours of Russian warships cruising in the Pacific Ocean.

Although there had been considerable discussion on the subject there were no fortifications at Port Phillip Heads.

Captain Gray, in the *Great Britain*, heralded his arrival in Hobson's Bay with a salute from his 'heavy deck guns'. A big ship arriving in a cloud of smoke amid sounds of gunfire was too much for the uneasy Melbourne population. Rumours spread like wildfire that the Russians had arrived.

There was a mass exodus for the hills. Gunsmiths did a roaring trade with those venturesome citizens who elected to stay and fight it out. The battle ended abruptly when port officials came ashore with news of the real identity of the invader. Captain Gray and the pilot were instructed to take more care when and where they permitted guns to be fired.

Commissioners from Victoria, New South Wales, South Australia and Tasmania met in Melbourne in August 1856 to determine a 'general and uniform scheme for the erection and maintenance of lighthouses on the Australian coasts'. After thirty-nine witnesses had given evidence, mostly master mariners and naval officers, the Commission recommended new lighthouses on Cape Borda and Cape Northumberland in South Australia; on Cape Schanck and Wilson's Promontory in Victoria; on King Island; and on Cape St George in New South Wales. Special reports were also made on harbour lights in each colony. Commander Kay RN, the Commissioner for Victoria, submitted that Gellibrand's Point light be improved; that the leading lights on Shortland's Bluff, although admirably sited, could be improved by the High Light being raised to enable it to be seen further away by ships approaching from the south and west, and over Point Nepean by ships approaching from the south-east.

These reports led to Cape Schanck lighthouse being completed in 1859 at a cost of £17,261; to the lightship being moored off Gellibrand's Point the following year; and to the new Shortland's Bluff leading lights in 1863. Lightships also marked the entrance to the Geelong channel, the northern entrance to West Channel. Swan Spit lightship was replaced by a lighthouse in fourteen feet of water in 1860. This structure was demolished in 1881 by the barque *Omeo* drifting over it in calm weather. It was replaced by another lightship, fifty feet in diameter and showing a red light from a height of thirty-eight feet, known as the Dancing Bottle.

Shifting sandbanks in and near West Channel have since resulted in the fairway being defined by a series of buoys, lighted beacons, and two leading lines. In 1930 Monash Light on the hill behind the quarantine area at Portsea and Wedge Beacon on Popes Eye Bank were established to lead ships through the outer reach. In 1947 Coles Channel Beacon was erected in a position where, in transit with Shortland's Bluff High Light, it leads shipping through the inner reach of West Channel. Coles Channel beacon, known as the White Lady, in transit with Monash Light is also the leading line for Coles Channel.

West Channel lightship was replaced by a pile beacon in 1881. Gellibrand Pile lighthouse was built in thirty feet of water, half a mile south of Gellibrand's Point in 1906 and replaced the lightship.

Wedge Beacon, named after John Helder Wedge the surveyor in John Batman's expedition, is a white hut on four piles in about eight feet of water. It is the only man-made structure known to be used for breeding by gannets. Since this peculiarity was discovered in 1966 it has been closely watched by wildlife observers. Gannets usually breed on remote islands and shun human company.

There has been a lifeboat stationed at Shortland's Bluff since the early 1850s when the pilots were instructed by Captain Ferguson to ensure that lifeboats were waterbourne at least once a month and fully equipped at all times. In 1856 an ordinary ship's lifeboat, without self-righting or emptying ability, was moored off the Doctor's Jetty and manned as required by the crews from the health officers' boat, the Customs boat and the pilots' boat.

Three years later a shed was erected on the pier to house a boat of improved pattern. It was first used in 1861 in an attempt to rescue the crew of the barque *Asa Packer,* wrecked on Nepean Reef in a gale. The lifeboat could not reach the wreck and the crew were eventually taken off by Captain Nicholson in the schooner *Empire.*

This lifeboat, and a smaller vessel kept at Point Lonsdale, attended mariners in distress for over thirty years, when it was replaced by a forty-five foot lifeboat built for the purpose at Williamstown. This was manned by a crew of nineteen, had two masts and three sails and carried two and a half tons of water ballast if necessary.

In 1929 the existing lifeboat – also forty-five feet long but powered by a diesel engine, named *Queenscliffe* – was built at Port Adelaide to an English design. Diagonally planked it has buoyancy tanks, 2 x 100 gallon fuel tanks, and a centre plate. The engine originally fitted was an eighty horsepower Wayburn petrol engine but this has been replaced by a seventy-two horsepower Gardner diesel. It is maintained and manned by the Ports and Harbours Department of the State Government.

Leading lights to guide ships safely between the sandbanks flanking the South Channel were established in 1873. The eastern, or high light, was erected on the foreshore at the foot of Arthur's Seat and exhibited a flashing light from an iron tower one hundred feet in height, with red sectors over the sandbanks and white sectors over the fairways. It is still known as Eastern Light but now shows an occulting light.

The western, or low light, was shown from a pile beacon in twenty-one feet of water at the eastern end of South Channel. It still shows a flashing light with white, red and green sectors and is known as South Channel Pile Light. In transit the two lights lead through the clear fairway of South Channel.

Until 1910 South Channel Pile Light was continuously manned by two men who also kept signals flying from a mast to indicate the state of the tide and to sound a fog signal when necessary. Around the inside of the gallery of the light are paintings of sailing ships, now badly eroded by the elements, which indicate how these lightkeepers spent their spare time. Fog and tidal signals from South Channel Pile light were discontinued on 6 January 1910. Since then the light has been unattended.

In 1903 a survey of the Rip indicated the possibility of deepening the entrance fairway to thirty-seven feet for a width of two thousand feet by blasting. The former Victorian gunboat *Albert,* three hundred and seventy tons, was used by the Ports and Harbours Department for blasting operations. On 27 May 1914, ss *Suevic,* drawing twenty-eight feet six inches, waited for seven hours outside the Heads until the tide rose to allow a depth of thirty-nine and a half feet over Rip Bank when she safely entered.

At this time a dredge from New Zealand had been chartered for two months deepening the eastern end of South Channel. Explosives used by the *Albert* were then stored at the explosives depot at Altona, near Williamstown, and valuable time was lost steaming up and down the bay to obtain supplies. When the South Channel Fort was abandoned as such, in 1914, the Ports and Harbours Department obtained permission to use the magazines to store explosives, for which purpose it is still used.

Another former gunboat, the four hundred and fifty-six ton *Paluma* from Queensland, replaced the *Albert* in 1916. Renamed *Rip* she was used for blasting until 1950 when the minesweeping corvette *Whyalla* was acquired, and also renamed *Rip*, by the Ports and Harbours Department. She was converted to their requirements, and although nearing the end of her useful economic life is still in service.

There were good reasons for the pilots' caution in taking deeply laden ships across Rip Bank. Probably the most serious incident was when the five thousand four hundred and forty-three ton ss *Cufic,* drawing twenty-seven feet eight inches, passed Point Lonsdale at 4.55 am on 21 September 1910 in the charge of Pilot Liley. It was the second hour of the ebb tide and the *Cufic* encountered a series of heavy seas which caused her to pitch heavily and ship water.

Neither the pilot nor the captain were aware that the ship had touched bottom, but after the pilot had disembarked the chief engineer reported to the captain that he had felt an impact. Other officers were questioned. Those stationed forward felt nothing, one who was aft thought the ship touched bottom. The second officer, who was in the crow's nest, felt three distinct bumps.

The ship was not leaking and the voyage continued. When the *Cufic* drydocked at Liverpool the shell plating was found to be set in, the stern frame fractured and the port propeller blades bent.

By 1916 the navigable depth in the entrance fairway had been increased to forty feet for a width of one thousand feet and work was continuing. From 1 February 1918 the pilots agreed to navigate vessels outwards provided they had absolute right of way over inward traffic.

When the First World War commenced in 1914 Port Phillip pilots were all commissioned as lieutenants in the naval reserve and undertook to control the movement of all merchant ships entering and leaving the Heads.

Armed with this authority the granting of right of way to outward bound deep draught ships presented no problems. But when compulsory examination ceased with the end of the war it became necessary to promulgate port regulations under which 'port closed to inwards traffic' and other local signals could be shown from Point Lonsdale signal station. Signals of this nature are still in use.

With the depth across Rip Bank increased to forty-three feet for a width of one thousand feet vessels drawing up to thirty-five feet could use the port. Blasting continued and, in 1973, there is a channel eight hundred feet wide with a depth of forty-eight feet, and forty-three feet of water for a width of one thousand four hundred feet. Vessels drawing up to thirty-eight feet can enter or leave safely, preferably entering on the flood tide. Because the deepest water is on the wrong side of the fairway for outward bound ships, deeply laden outward bound vessels are given absolute right of way, in order to allow them more room to manoeuvre.

To take full advantage of the depth available at the Heads, plans were made to deepen the channels towards the Williamstown and Geelong oil terminals. Shell Oil Company made special arrangements for their seventy thousand ton tanker *Philine* to make a series of trial runs across Rip Bank, ballasted to a draught of forty-two feet. Special equipment ashore and afloat measured scend and the effect of sea and swell to determine the safe maximum at which tankers of this size could use the port.

Before the results of the experiment became effective oil was discovered in Bass Strait. The oil companies saw the possibility, which has since materialised, of a large proportion of Victoria's oil needs coming by overland pipeline. Meanwhile the plans for deepening the channels specifically for tankers were shelved.

Progressive deepening of the entrance fairway demanded improved leading lights on Shortland's Bluff. In 1918 another light was shown from an obelisk east of the Low Light. This new light, in transit with the High Light, marks the eastern limit of the forty-three foot channel across Rip Bank.

In 1924 Hume Tower was built westwards of the Low Light. The light shown from this tower, in transit with the High light, marks the western limit of the thirty-seven feet channel across Rip Bank. Two white diamond-shaped beacons, unlighted, when in transit, mark the eastern limit of the forty-eight feet fairway. High Light one quarter of the distance between Low Light and the obelisk light is the night mark.

Originally the maximum depth at the eastern end of South Channel was twenty-seven feet and this was necessarily increased as the depth in the entrance fairway was improved. In January 1909 the dredge *John Nimmo*, built for the purpose but previously used to dredge Hopetoun Channel into Geelong, commenced deepening what has since been known as Dredged Cut, to a depth of thirty-one feet for a width of four hundred feet over a distance of one and a half miles. This cut, parallel to the South Channel leading line but north of the South Channel Pile light, has since been progressively deepened until there is a depth of forty-three feet for a width of seven hundred and fifty feet over a distance of six thousand seven hundred feet.

Any vessel able to enter or leave over Rip Bank is able to pass through South Channel.

Since 1924 the eastern extremity of Middle Ground sandbank, where shipping turns into or out of the South Channel, has been marked by a pile lighthouse known as Hovell Pile Beacon.

CHAPTER 6

Boats in Trouble

Shipwrights have been employed by the pilots at Queenscliff since the earliest days of the cruising pilot ships. Besides maintaining the schooners in first class condition work boats were built and kept in good repair. These boats, working in all weathers at all hours of the day and night, had to be of the highest standard and were constantly exposed to damage.

Over a century of experience in a craft, handed down from father to son, has developed boats that are probably the best for their purpose anywhere in the world. Eighteen feet long by six feet beam, clinker built and single-ended, they have always been slung from davits on the cruising pilot ships and lowered into the water at every operation of embarking or disembarking a pilot. Except for the fact that oars and sails have given way to petrol, and more recently to diesel engines; that manhandled rope tackles have been replaced by electric winches and wire falls; the boats at present in use are almost identical to those used by the sailing cutters.

Bitter experience of tragedy and disaster has, of necessity, developed these boats to their high state of efficiency. There have been several minor collisions involving the cruising pilot vessels and ships taking pilots but few accidents have occurred during the transfer of pilots.

The *Geelong Advertiser* reported in 1854 that the ship *Flavius* ran down the cutter *Corsair* with such force that six pilots escaped by jumping on to the ship. Shortly after the *Rip* was commissioned she was run down by ss *Alhambra,* and in the early seventies was again in a collision with the *Palm Tree,* sustaining such damage as to necessitate the schooner's forepart being rebuilt to the foremast.

There have also been several accidents to boats communicating between pilot ship and the shore. On 25 March 1886 a seaman was drowned when one of the *Rip's* boats was swamped close to Point Lonsdale lighthouse. In those days the schooners hove to near the lighthouse while a boat went ashore for mail and other communications. On this occasion the boat was towed close inshore by the inward bound tug *Rescue.* Breaking seas near the boat channel deterred the two seamen in the boat from attempting the landing, and while endeavouring to return to the schooner the boat was swamped and capsized. Although buoyancy tanks kept the boat afloat one of the men lost his hold and was swept to his death in the boiling surf. Some fishermen eventually secured the upturned boat and rescued the survivor.

There was a similar tragedy in August 1927 when a boat left the anchored pilot steamer *Victoria* to pick up Pilot Banks at Queenscliff. Tide was ebbing strongly and the night was dark. Unable to make progress against the tide under oars or sails, the two seamen were again trying to get back to the pilot ship when their boat was swept seawards. Their shipmates aboard the *Victoria* were unaware that the boat had not reached the shore safely. In the swirling Rip the boat capsized, lost oars and sails, but the two men hung on until another sea turned it right side up again. Throughout the cold stormy night the two men crouched in the boat with no food or means of propulsion, fighting for survival.

Meanwhile, aboard the *Victoria,* it was not known that the boat was missing until the outside steamer, *Akuna,* came in to Queenscliff to investigate the delay. Both steamers, and the Queenscliff lifeboat, then searched all night with no avail.

During the night the elder of the two seamen in the boat died of exposure and the survivor drifted helplessly in the waterlogged boat. After an ordeal lasting thirteen hours the boat was cast ashore near Sorrento and he collapsed on the beach. He was eventually found wandering in the scrub near Rye.

CHAPTER 7

Ships in Distress

But ships were more often in trouble outside the Heads than were boats. In 1861, soon after Pilot Singleton had disembarked, the outward bound schooner *Lilies* was dismasted in a southerly buster. The *Corsair* succeeded in towing the disabled schooner to an anchorage and was awarded £500 salvage.

Pilot Blanchard was in charge of the barque *Craigburn,* being towed seawards through the Rip by the tug *Rescue* on 8 May 1891. The tide was full ebb and a strong southwester raised the usual turbulent seas. Almost clear of Point Lonsdale the towline accidentally slipped. Sails were hurriedly set in an effort to claw offshore but twice the barque missed stays and drifted towards the lee shore.

Blanchard ordered both anchors let go and signalled the master of the *Rescue* to return. The tugmaster replied that his towage contract expired when the towline was slipped and demanded another £500 or a salvage agreement. Relying on the *Craigburn's* anchors the captain disregarded the pilot's advice and refused the tugmaster's demand. Two other towage offers, from the *Racer* and the *Eagle* were also refused. Blanchard eventually took the initiative and ordered the barque's crew to take the tug's line.

Two sailors forcibly restrained the still resisting captain from interfering. *Eagle's* towline was accepted but the carpenter refused to raise the anchors or slip the cables. *Eagle* commenced towing but parted the towline against the weight of the anchors. *Racer* passed a towrope but this also parted. Then the anchors dragged and *Craigburn* fired distress rockets.

At 8 pm the Queenscliff lifeboat was unable to proceed against the flood tide and was eventually towed out through the Heads at 11 pm. By then increasing wind and seas had decided the pilot to order the *Craigburn* abandoned. With fourteen men in the starboard lifeboat he was picked up the following morning.

The captain and fifteen men elected to stay aboard the barque. When the *Craigburn* was in the breakers twelve more men left in another lifeboat which immediately capsized and only seven men reached the shore alive.

Only the captain, first mate, carpenter and one seaman were left aboard the doomed *Craigburn,* on the rocks about four hundred yards offshore near Sorrento. Although the wind had abated there was a heavy surf, which prevented a seawards approach by the lifeboat.

Rocket lifesaving gear arrived on the beach at 2.30 am. At daylight the hawser was connected to the mainmast. First the captain was hauled ashore in the breeches buoy, then the carpenter. The remaining seaman was seen grappling with the first mate, who was apparently hopelessly drunk. Almost too late the seaman abandoned the rescue attempt and it took sixteen hazardous minutes for the shore party to haul him through the wreckage-strewn surf to safety.

There was a happier ending when the ship *Loch Etive* was sighted from the pilot schooner *Hawk* close to the breakers off Barwon Heads on Christmas Day, 1892. After the pilot vessel had signalled 'Haul your wind on the starboard tack', Pilot McWilliam was put aboard in water too shallow for the *Hawk* to approach in the usual manner. With a fresh south-wester and the flood tide *Loch Etive* was safely extricated.

Only five days later the *Loch Vennacher* embarked a pilot two hundred and sixty-nine days out from Glasgow. Soon after the anchor was down in Hobson's Bay the crew assembled to give three hearty cheers for the captain and officers in appreciation of their safe arrival after the danger and difficulties encountered on the voyage.

While men were aloft furling sails in mid-ocean the captain observed a series of great sea waves hurtling towards the ship and hailed those aloft to come down. As the men reached the deck the *Loch Vennechar* rode safely over the first freak wave and dived into the trough as the second sea struck. The crest curled over sixty feet above the deck, filled the lower topsails, and a solid mass of water swept the ship from stem to stern.

Spilling from the decks the receding water revealed fore and mainmasts broken off near the deck and the mizzentopmast gone. Under jury rig the *Loch Vennachar* put into Mauritius for repairs. Ten years later she had to be beached after a collision in the Thames, and in September 1905, the *Loch Vennacher* was lost with all hands on a voyage to Adelaide. Wreckage found washed up in what has since been known as Vennachar Bay, on Kangaroo Island, indicated that she had met a similar fate to the *Loch Sloy* six years earlier.

The Italian barque *Lothair's* master mistook Barwon Heads for Port Phillip entrance when he was heading for Melbourne for repairs on 20 April 1893; while on a voyage from Hong Kong to Callao. When the *Lothair* was sighted off Split Point with the signal flying for a pilot the captain reported that the barque would only sail on the port tack because the rudder was twisted. The tug *Eagle* towed the Queenscliff lifeboat to her assistance and later towed the barque to Melbourne.

After being repaired the *Lothair* was restrained from resuming her voyage because the towage account of £1050, and port dues of £300, were outstanding. Eventually the captain was forced to sell some of his cargo of tea, rice, silk and Chinese curios to liquidate the debt before clearance was granted.

After twenty years in the Pilot Service, Pilot Press was severely reprimanded over the loss of the Asiatic Petroleum Company's ss *Petriana* on 28 November 1903. When she arrived on the pilot station with a cargo of petroleum products for Melbourne Pilot Press was embarked about four and a half miles south of Point Lonsdale in misty weather at 6.30 am. The flood tide was making and the *Petriana's* maximum speed of five and a half knots was called for as a course was set through the entrance.

Fog obscured the Heads at 7 am but Arthur's Seat could be seen and Point Lonsdale fog signal was audible. At 7.13 am the pilot noted the bearing of the fog signal changing rapidly, assumed the ship was feeling the full strength of the tide and altered course westwards with the intention of anchoring until visibility improved.

But the *Petriana* touched bottom as he gave the order and the flood tide pinned her to the rocks. The tug *James Paterson,* hastily summoned to her assistance, failed to refloat her. Next day the flood tide swung the ship's stern until she parallelled the beach and the next ebb tide turned her again, grinding the bottom into the rocks seawards of Point Nepean; since known as Petriana Rocks.

Hard aground and holed amidships, engine and boiler room flooded, salvage attempts were abandoned on 30 November. Most of the bulk oil cargo had been pumped overboard in an effort to lighten the ship. This created Victoria's first oil pollution problem and fish caught in the vicinity were said to have had an oily taste for months afterwards.

Twenty-seven Asiatic crew members went to Melbourne in the *James Paterson* and presented another problem: the recently enacted Immigration Restriction Act forbade them to set foot on Commonwealth territory and they were detained aboard the tug. Hire of the *James Paterson* cost £7 an hour. Fortunately the problem was solved by the Japanese ship leaving the following day.

Probably the most spectacular wreck at Port Phillip Heads was that of the six thousand nine hundred and one ton P and O liner *Australia* on 20 July 1904, in light winds and moderate seas. Two years previously, when the existing lighthouse on Point Lonsdale was completed, the character of the light was changed from green with red sectors over the dangerous water to white with red sectors. Altering course too soon after clearing the red sector over Corsair Rock – particularly when entering against an ebb tide – has been the cause of many steamships striking Corsair Rock. After altering course the tide is athwart the vessel's track and setting her strongly towards the adjacent Nepean Reef.

Pilot Dennis, with more than twenty years service to his credit, boarded the *Australia* at 1.20 am, four and a half miles from Point Lonsdale. Ordering full speed when he arrived on the bridge he noted that the leading lights were open slightly to the west. The flood tide would set the ship further westwards and accelerate the ship's fourteen knot capability to about twenty knots. To bring the leading lights into line Dennis ordered the course altered to starboard. The leading lights were not in line by the time Point Lonsdale light turned red and the pilot ordered the helmsman to steer more to starboard.

Captain Cole, the *Australia's* master, heard the orders, saw the leading lights open to the east, and then the dark land mass of Point Nepean come into view as the ship continued the swing to starboard. The pilot ordered 'Steady', the quartermaster gave the course 'N80°E'. Point Lonsdale light still flashed red.

Within a minute the *Australia* ground to a halt on Nepean Reef. Fortunately only forty-eight of the four hundred passenger berths were occupied but there was a crew of two hundred and forty-six on board; and two thousand one hundred and seventy-six tons of general cargo valued at £25,000. With her bottom ripped open for half her length it was soon obvious that the *Australia* would not be refloated. Passengers and crew were landed at Queenscliff by the ship's lifeboats without incident.

Pilot Dennis collapsed soon after the stranding. The next day he was found to be a victim of Bright's disease, and its associated disorders, including defective eyesight.

At the Court of Marine Inquiry, which decided that the wreck had been caused by his wrongful act, he claimed he could not clearly make out the leading lights when he altered course. These disclosures awakened the Marine Board of the necessity for pilots to undergo more stringent eyesight and medical examinations. Since then pilots under the age of fifty have been medically examined on alternate years; those over fifty every year. All pilots must have their eyesight tested every year.

The wreck of the *Australia,* which experts declared would not survive the next south-westerly gale, was sold at auction for £290, the cargo for £60. It was two years before the wreck broke up, during which the enterprising purchaser made a fortune from salvage operations. The last vestige of the wreck was blasted apart in 1911 because it was blocking a channel used by fishing boats.

There was another tragedy in the pilot service on 24 February 1909 when the outward bound barque *Gladys* anchored off Shortland's Bluff, and the captain reported Pilot Stanford missing. When the pilot was last seen at 4 am the barque was off Schnapper Point. The helmsman said he had heard a splash about fifteen minutes later. *James Paterson* was towing the *Gladys* and it is likely that Stanford toppled overboard while standing on the taffrail to observe the tug.

Captain Watson, of the *James Paterson*, reported passing Gellibrand Light at 1.30 am, the tug five hundred feet ahead of the barque. Hazy weather was encountered approaching South Channel and at 5.30 am Captain Watson, in the puzzling absence of orders from the pilot, turned tug and tow round to await improved visibility. He resumed the tow half an hour later and was off Queenscliff at 7 am when hailed from the *Gladys* with the information that the pilot was missing.

The barque's captain had left the deck at 3 am. The helmsman, a lookout and a boy were keeping watch with the pilot. The lookout last saw the pilot when he went forward just before 4 am and remarked about the haze.

Frederick Henry Sanford, a native of Yarmouth, England, who had celebrated his sixty-second birthday the previous Saturday, was master in Howard Smith's ships and wharf manager for that company at Sydney before joining the Pilot Service in 1889.

CHAPTER 8

From Sail to Steam

Captain William Howard Smith, chairman of the 1886 Board of Inquiry into the working of the pilot service, was assisted by another shipmaster turned shipowner, Captain Archibald Currie; and a Union company master, Captain Thomas Underwood, exempt from taking a pilot at Port Phillip, was the third member of the Board. Several pilots, former pilots, shipmasters and other shipping identities, gave evidence before the Inquiry, which lasted from 11 to 30 November.

The Board then decided that

> The Sea Pilots of Victoria numbered for many years past twenty-two, but are now twenty-four, an increase of two having recently been made by the Pilot Board to meet the increased trade of the port.
>
> These twenty-four pilots are divided into two companies, called respectively No 1 Company and No 2 Company. Their functions are to pilot all vessels from sea not commanded by exempt masters to a berth alongside one of the piers or wharves in Hobson's Bay.
>
> Actual earnings of the pilots have for many years been veiled in mystery.
>
> The 'outside station', or the time allowed for each company's schooner to remain outside Port Phillip Heads, giving the company to which it belongs the monopoly of supplying all inward bound ships with pilots, is eleven days. To the company securing the inwards pilotage of a ship is also assured the right of outwards pilotage.

When the pilot schooner takes up her station six pilots are required to be on board. When, at any time, the pilots on the outside schooner are exhausted the schooner lying at anchor at Queenscliff is allowed the privilege of going outside and supplying inward bound vessels with pilots; but, as soon as any other pilots belonging to the company which has the outside station arrive, this privilege ceases.

The outside schooner is bound to take off from all outward bound vessels, not only the pilots belonging to their company, but also the pilots belonging to the other company.

It is occasionally the custom for the outside schooner to leave her station and send a boat to Point Lonsdale with or for pilots. In our opinion we consider this practice highly objectionable – that the outside pilot vessel should ever leave her station at any time.

Regarding the appointment of pilots we noticed the following anomaly: that some candidates who were comparative strangers to the port were appointed pilots upon their first application in preference to others who were of well-known local standing and high qualifications, and who had made several previous attempts to obtain the position of pilot without success.

We consider twenty-four pilots quite sufficient but recommend that, in future, ten should reside at Queenscliff and fourteen at Williamstown.

One boat is not sufficient to supply the wants of all inward and outward bound ships when a rush of shipping takes place. In view of the age of the *Rip* [twenty-seven years] and the *Mavis* [thirty-four years] we recommend that the pilots purchase another first class vessel.

There is no real competition at present, it being now merely a service of payment by results; but if the above recommendations are carried out it would stimulate the pilots to greater exertion and make the service as competitive as the present system will permit.

We recommend the removal of the Pilot Office from Williamstown to Melbourne; that the present notice of

> twelve hours be dispensed with, so that, when a pilot's services are required he should at once proceed on board the vessel requiring him.
>
> No pilot should be allowed to hold any municipal, volunteer, or any other appointment or employment other than that connected with the Pilot Service.

These recommendations so ludicrously favoured the shipowner that it is not surprising that they were not wholly enforced.

It was the disclosures of the system appointing pilots that led to the present requirement for an applicant to hold a Foreign-Going Master's Certificate, and have one year's experience in command while exempt from pilotage at Port Phillip.

Between 1854 and 1876 the few vacancies in the Pilot Service had been filled by the mates of the pilot schooners, who had accepted such employment as 'apprentice pilots'. After 1876, as the veterans who established the service in the 1850s retired, vacancies had become too frequent to be satisfactorily filled by pilot schooner mates. Applications were sought further afield.

Analysis of the records show that, during the ten years preceding the 1886 inquiry, twenty vacancies had occurred. First of these was filled by J. B. Maynard, an American-born mate of the pilot schooner, who does not appear to have been called upon to substantiate his claim to have been in command of a fifteen hundred ton American ship. Four local shipmasters unsuccessfully applied.

Fourteen sought the next vacancy the following month. Successful applicant was C. R. Devlin, another mate of the pilot schooner who had previous experience as master in steam and sailing ships. Then, in 1878, a Melbourne piermaster, G. F. Dennis, with previous experience as mate in foreign-going ships, was appointed from the twelve applicants, including two pilot schooner mates and five shipmasters, ranging in age from thirty-one to fifty-seven.

Only two weeks later the thirty-three year old mate of the *Rip*, George Bevis, was selected from eight applicants. Then, in

March 1879, thirty-two year old Richard Dawes, master of the brig *Alexander,* the youngest of twelve who applied, succeeded. He was followed soon afterwards by John Neilson, thirty-eight year old master of the coastal steamer *Julia Percy,* making his fifth application.

Eleven applied again early in 1881, and John Cairns, another local shipmaster, succeeded at his fourth attempt at the age of thirty-six. Six months later the eldest applicant, thirty-nine year old John Stalker, formerly in command of a New Zealand government steamship and of the brigantine *Vistula,* presently serving in HMVS *Cerberus,* succeeded with his second application. Edward Thurnall, formerly a shipmaster now mate of the *Rip* – at the age of twenty-eight the youngest of his nine contenders – was selected the same year.

But the tendency towards younger men was reversed in May 1883 when the sixteen applicants included a twenty-one year old, the twenty-six year old mate of the *Rip,* and a forty year old shipmaster. Henry Press thirty-nine year old master of an ocean-going yacht was successful. Nine of these applicants tried again four months later, with nine others, whose ages ranged from thirty to forty-four. John Stewart, aged thirty-five, and master of the *Cingalese,* succeeded. There was another opportunity five weeks later when a master, thirty-four year old Edward Clayton, was appointed but died ten months later.

This vacancy was then filled by Francis Burton, thirty-nine year old master of the British barque *Loch Katrine,* who was selected from seventeen applicants at his second attempt. Then, some time later, thirty-four year old Alexander McWilliam, master of the barque *Glenshee* succeeded at his first try; followed by thirty-three year old Alexander McIntosh, master of ss *City of Adelaide,* selected from eighteen applicants, at his third attempt.

George Service, master of Archibald Currie's ss *Bucephalus,* and formerly in command of large sailing ships, was, at the age of forty, selected from seventeen applicants in December 1884; followed by J. A. Mitchell, master in sail and steam at the age of thirty-one, and youngest of sixteen who applied. Captain Bennett, well known master of the ill-fated *Loch Vennacher* was an unsuccessful thirty-nine year old applicant on these two occasions.

In April 1886 William Emmerson, thirty-four year old steamship master, was selected from fourteen applicants; and there were twenty applicants between the ages of twenty-six and forty-two the following month when thirty-five year old Daniel Russell, Union Steamship Company master, and another local shipmaster, thirty-seven year old Richard Wills were both appointed.

Anticipating Captain Howard Smith's sympathy, and disgruntled by the method of selection, six unsuccessful applicants wrote to the Board of Inquiry:

> . . . the stringent regulations lately framed by the Victorian Pilot Board regarding the admission of candidates . . . the age limit is forty next birthday . . . the rule for applicants to have one year's command, or three years as mate of a square-rigged vessel, prevents us from ever becoming pilots . . . almost an entire stranger, a master of a square-rigged vessel who has no experience in steam has been appointed . . . some of the regulations seem purposely intended to favour the class of men who never possessed an exempt master's certificate; and, to enable them to perform their duties, the position of leadsman was introduced . . .

Two of the signatories to this letter, thirty-seven year old Frederick Strickland and thirty-eight year old Frederick Stanford, masters in Howard Smith's steamships, successfully applied in 1890. By then the Marine Board had taken over the functions of the Pilot Board. But experience in square-rigged ships continued to be a qualification for applicants. Pilot Colin Goss, who joined the service in 1947 and retired in 1967, was the last to hold this qualification.

Henry Draper, who became a pilot in 1852 and was the last of the foundation pilots, retired in 1889. With his retirement there had been a complete change in the complement of the service since 1876. By 1890 the pilots would have been almost equally divided between sail-trained and steam-trained men; gradually eroding the solid majority of the older dyed-in-the-

Hobson's Bay in the 1870s

wool sail-trained pilots who had unanimously favoured continuing with sailing schooners.

With the *Rip* and the *Mavis* approaching the end of their useful economic life there was a distinct strengthening of opinion within the service to acquire a steam pilot vessel. During the 1890s negotiations with the government resulted in the promise of a loan of £15,000, principal and interest repayable over twenty years, for the purpose. This was the current estimated cost of building a steam pilot vessel.

Extensive discussion among the pilots resulted in a vessel one hundred and forty feet long by twenty-three feet beam, being decided upon. Single screw propulsion with a nominal horsepower under one hundred was desirable, for economic reasons.

Late in 1898 Pilots Barber, Stalker, McIntosh, Mitchell and Wills were selected as the 'Steam Pilot Boat Committee' empowered to engage a naval architect, decide upon the exact requirements, prepare plans and specifications, and then call for tenders. Pilot Stalker was elected chairman and Pilot Mitchell honorary secretary, positions which they retained throughout the existence of the Steam Pilot Boat Committee.

Mr Findlay, a Williamstown domiciled engineer who had offered his services, was retained as consulting engineer and expert adviser. He suggested two water tube boilers, which would make three-quarters of maximum power available from either through triple expansion reciprocating engines. Findlay considered fourteen knots could be maintained on nineteen tons of coal a day, but that three tons a day would maintain a cruising station. In his opinion competent builders could be found in Victoria but might charge up to fifty per cent more than United Kingdom shipbuilders.

Approaches were made to the government with a suggestion that a subsidy be granted and a more powerful vessel be built which would be available for defence purposes in the event of war. Although the subsidy could not be granted the pilots were requested to submit the tenders to the government before a contract was signed.

Plans and specifications were ready by the end of 1898. Soon after calling for tenders the Steam Pilot Boat Committee were told of a twin screw steam yacht, one hundred and forty-five and a half feet by twenty-three feet, built in 1891 and capable of fourteen and a quarter knots, available for £13,000.

The next development was at the end of January 1899, when a deputation from the Trades Hall Council, accompanied by three Labour Members of Parliament, made representations to the Committee requesting a clause in the contract that the steam pilot vessel be built in Victoria. Besides providing jobs for Australian workers, the deputation submitted that funds provided by the Victorian government should be spent in the colony. Acknowledging cost as an important factor they thought the local price might compare favourably with overseas estimates when cost of delivery was added.

Three tenders, ranging from £17,900 to £19,000, were received from Britain. Victorian shipbuilders submitted tenders for £21,372. Cost of delivery from Britain was estimated at £1300.

Because all the tenders exceeded the promised loan of £15,000 another approach had to be made to the government for an additional loan. Purchase of the steam yacht was suggested as an alternative, but not recommended because of the reduced security to the lender and less value to the borrower. Sir George Turner, the Victorian Treasurer, queried the higher estimates but accepted the explanation that shipbuilding costs had increased by thirty per cent in two years.

On 17 April the tender from the Australian Forge and Engineering Company of Melbourne, to build a steam pilot vessel for £21,905, with delivery in fifteen months, was accepted subject to conditions imposed by the Commissioner of Trade and Customs. An additional loan of £8000 was made available, principal and interest amounting to £1250/10/10 to be repaid annually for thirty years commencing on 30 June. £23,000 was paid by the Treasury to the Port Phillip Pilots Steam Pilot Boat Committee on 15 September 1899.

Victoria was selected as the name for the new ship 'in honour of our Queen and Colony'. But a request for a portrait of the Queen for display in a prominent position on board, and for permission to fly the Blue Ensign, were refused by Buckingham Palace.

Point Lonsdale Lighthouse

The diesel-electric pilot vessel Wyuna

Captain J. G. Caught pilot 1851-1880

Pilot steamer Akuna II, formerly HMAS Gladstone

When the contract time expired on 15 May 1900 *Victoria* was far from complete. The builders agreed to forfeit £40 a month during the extended time for completion, estimated as the end of October. She was launched on 11 September, and collided with ss *Maitland,* damaging the new ship's rudder post, and further delaying delivery.

However, Notices to Mariners were circulated on 26 September that a steam pilot vessel would be cruising off Port Phillip Heads from the beginning of 1901. The shipbuilders were advised that if *Victoria* was not available then cost of chartering another steamship would be deducted from the contract price.

SS *Albert,* operated by the Ports and Harbours Department and owned by the Victorian government, acted as a pilot ship until *Victoria* took up her station on 19 January, and again between 2 and 11 February when the new ship had to return to Williamstown because the evaporator was unsatisfactory. She was destined to give little more serious trouble during fifty-six years of service as a pilot steamer.

Final settlement of the building contract was negotiated on 30 September for £23,500. One month later, after completing their task to the satisfaction of all parties. the Steam Pilot Boat Committee was dissolved.

With the commissioning of *Victoria* the schooners *Rip* and *Mavis* were sold. *Hawk,* the remaining schooner, was retained to relieve *Victoria* on the outside station. But the Notice to Mariners had clearly stated that a steam pilot vessel would maintain a continuous cruising station off Port Phillip Heads. It was obvious that a second steamship would have to be obtained.

Early in 1903 another loan was negotiated to purchase the steam yacht *Alvina* for £11,200. This one hundred and ninety-four ton luxuriously appointed vessel, built at Southampton, England, in 1887, was one hundred and thirty-nine feet long by twenty feet beam, and propelled by triple expansion engines and a single screw. Between 1891 and 1899 her owner was Laurence Cave, for whom she flew the burgee of the exclusive Royal Yacht Squadron, and the White Ensign.

HRH the Prince of Wales, later King Edward VII, was entertained on board; as was Lily Langtree, the famous actress. In fact *Alvina's* elegant appearance, and the association with the actress, was said to be reason for her unofficial name *White Lily.* Marmaduke Fox, a Yorkshire woollen millionaire, bought *Alvina* in 1899, and in 1902 she was laid up at Glasgow under the registered ownership of G. W. Wallace of London. She left the Clyde on 12 April 1903 bound for Melbourne via Suez with a delivery crew, and arrived in Hobson's Bay seventy-three days later.

Soon afterwards she alternated with *Victoria* on the outside station at weekly intervals. While the outside ship maintained the cruising station the inside ship replenished bunkers, fresh water and supplies, and accommodated pilots while anchored off Queenscliff.

By this time the twenty-four pilots were operating as a single company, pooling their resources and dividing the monthly takings after expenses had been paid. Monthly instalments of principal and interest were also being repaid to the Treasury.

Each new pilot acquired a financial interest in the ships and ancillary property, this 'ingoing' being paid to the retiring pilot to relinquish his interest. This agreement is in the form of the Pilot's Financial Rules, a Deed signed by each pilot on payment of his 'ingoing'. These Rules are altered from time to time to prescribe the sum which new pilots pay for their share in the assets, the figure retiring pilots receive.

Port Phillip Pilots were never incorporated and have no legal identity except as individuals. Nevertheless, collectively, it was incumbent upon them to provide for depreciation of assets, and to budget for replacements at escalating costs.

Provision was made for this in 1907, when a Reserve Fund for Extraordinary Expenditure was established by monthly contributions from earnings. Every pilot had to sign the Deed establishing this fund, and one of the first duties of new pilots is to sign in confirmation. This fund has been used to pay for repairs to pilot ships and launches, occasionally for income tax and legal commitments, and for accountancy charges. Its present principal function is to pay 'outgoing' to retiring pilots.

In cases where a new pilot replaces a retiring pilot this necessity does not arise. But it has happened that new appointments have not been made to replace retiring pilots and the 'outgoing' has been provided from the Reserve Fund.

In 1934 it was deemed necessary to limit the Reserve Fund and channel the contributions to a 'Salary Equalisation Fund' established by separate deed. If in any month there was more money than required to meet the pilot's remuneration, the balance was paid into the new fund as a reserve in months when there was insufficient income. Then, in 1937, the Salary Equalisation Fund was limited by yet another deed to £5000, anything surplus being paid into a Ship Replacement Fund in anticipation of the imminent replacement of the two steam pilot vessels, which had then seen more than thirty years service.

Alvina had been replaced in 1925 by a nine hundred and fifty-three ton twin-screw steam yacht built for the German Administrator of New Guinea in 1911 and named *Komet*. When the 1914 war started *Komet* was commissioned by the German Navy as a patrol ship in New Guinea, and captured soon afterwards by the Royal Australian Navy. Because she was the first enemy vessel to be captured and recommissioned by the RAN she was given the name HMAS *Una,* and used for patrolling the islands north of Australia.

HMAS *Una* was withdrawn from commission in 1924 and laid up at Sydney. The following year she was bought by Port Phillip Pilots and converted into a cruising pilot ship. *Akuna* was selected as a suitable name, not only because of its aboriginal description of 'flowing waters', but also because *Una* was already painted on stem and stern and only needed the prefix AK. In 1925 this magnificent vessel, two hundred and ten feet long by thirty-one feet beam, joined *Victoria* on the cruising station.

Alvina was sold and converted into a ferry in Westernport, and later into a floating casino within Port Phillip. As such she attracted police attention, was raided three times, and eventually laid up. She was recommissioned in 1943 to carry out salvage operations on the sunken wreck of the *Kakaraki* in Hobson's Bay; for fishing expeditions, and finally in the Bass Strait cargo trade.

She was driven ashore in a gale, while anchored off Stanley,

Tasmania, in 1951, and refloated by local fishermen. Renamed *Tamar* her last voyage was to Launceston, where she was used as a training ship for sea cadets until broken up in 1961.

Shipping, and the earnings of the pilots, fell to a low ebb during the Second World War, and the two cruising ships proved too expensive to maintain. The larger and more expensive *Akuna* was laid up at Geelong in 1941. *Stradbroke II,* a motor yacht bought for £8000, then acted as relief ship for *Victoria* on the outside station; but this arrangement proved unsatisfactory and as soon as shipping increased *Akuna* was recommissioned. By the time the war ended both ships were showing their age. Another 'New Ship Committee' was formed within the Pilot Service to investigate replacements.

London's Trinity House, and several United States ports where cruising pilot ships were maintained, were written to with a view to purchasing suitable ships. None were offering.

CHAPTER 9

Born 1839, Still Going Strong

Early in 1948 negotiations were initiated with the State government for a loan to build a new ship. Agreement was reached on 14 December 1949 and Mr C. S. MacKinnon was engaged as consulting naval architect. After protracted discussion and modification, plans and specifications were finalised in July 1950.

Australian, British, Dutch and Hong Kong ship-builders all submitted tenders ranging from £A304,572 to £A420,000 from British yards; £A431,396 from Holland; £A437,390 from Hong Kong; and £A704,000 from Sydney. In January 1951 a contract was signed with Ferguson Brothers of Port Glasgow to build the ship for £A362,695, with delivery in twenty-one months.

Wyuna was selected as the name for what is claimed to be the best cruising pilot ship afloat; one thousand three hundred and four tons, two hundred and eight feet long by thirty-nine feet beam, twin screw diesel-electric propulsion operated by automatic bridge control. Power for a maximum speed of fourteen knots is supplied by three by six hundred and fifty horsepower English Electric diesels.

Four months after being launched on 15 April 1953 *Wyuna* was taken over by a delivery crew and arrived at Williamstown on 13 October. One month later she commenced her first cruising station. Total cost was in the vicinity of £A480,000. Fuel capacity is sufficient for *Wyuna* to remain on station for twelve months without refuelling, and it was unnecessary to maintain two cruising pilot ships in commission. *Akuna* was sold to shipbreakers and *Victoria* maintained in a state of readiness at Williamstown to relieve on the outside station at six-monthly intervals.

Two crews were kept on the payroll on a week on, week off basis, permission being granted by the Marine Board for *Wyuna* to come off station between 9 am and 11 am every Monday to change crews.

But *Victoria* was now more than fifty years old, and expensive to maintain even as a relief ship. In 1955 another 'Replacement Ship Committee' was formed to investigate several vessels then offering for sale. The Second World War minesweeping corvette HMAS *Gladstone* was selected. Cost of conversion was estimated to be in the vicinity of £37,000.

Renamed *Akuna II* the converted minesweeper took over relieving pilot ship duties from the fifty-five year old *Victoria* in 1957, when the veteran ended her days in the shipbreaker's yard on the Maribyrnong River.

All over the world a tendency developed to replace cruising pilot ships with high speed motor launches. When an oil refinery was established at Westernport in 1966 Port Phillip Pilots undertook to supply pilots to tankers bound thereto from the Port Phillip cruising pilot station. It soon transpired that high speed launches could be used more effectively from Westernport.

Two fast launches similar to those used by Trinity House pilots, and other pilotage authorities, were acquired by the Port Phillip Pilots and named *Hawk II* and *Alvina II*. They are forty feet in length, powered by twin screw Cummins diesels and capable of a maximum speed of seventeen knots. Both have identical fibreglass hulls. *Hawk* was imported complete from the United Kingdom; *Alvina's* hull was imported and she was fitted out by the pilots shipwrights at Queenscliff.

Then, in 1973, Pilots Lewis and Elford went on a tour of United States and European ports to investigate the use of similar launches instead of replacing the ageing *Akuna II* with a cruising pilot ship. Their report is not yet to hand but a third launch has been ordered in Britain. There is a possibility that one hundred and twenty-two years of pilot ships continuously cruising off Port Phillip Heads may be nearing its end.

Although this may eventuate as a relieving system, using Westernport as an alternative when the Rip is untenable for launches, the peculiarities of that notorious stretch of water render a round-the-clock pilot service by launches from Queenscliff unlikely.

Circumstances that demanded a cruising pilot station in 1851 still exist: masters of ships requiring a pilot will find themselves in the dangerous tidal streams and turbulence of the Rip if a conspicuous pilot vessel is not maintaining a station several miles offshore. And the Rip, particularly during ebb tides, is not a safe area in which to continuously operate forty foot launches.

While cruising off Port Phillip Heads pilot ships have frequently been involved in situations not concerned with piloting. One of these was in February 1960, when seventy-four commandos from an army unit embarked on an ill-fated training exercise intended to be a simulated raid on the Army Officers Training College at Portsea. Amphibian ducks, kayaks and rubber dinghies were used by the raiders to cross the unpredictable tidal streams between Queenscliff and Portsea.

The exercise was timed to commence just before the ebb tide made at 6 pm. All the commandos had strict orders to keep in close proximity to their escorting army towboat. But a combination of the towboat breaking down, deteriorating weather and an early ebb tide, caused the entire expedition to be swept seawards into the usual turbulence of the Rip during an ebb tide. Three commandos were drowned when their duck capsized and sank in the Rip.

Fortunately the lighthouse keeper at Point Lonsdale, keeping a close watch on the situation, anticipated the danger and alerted *Akuna* on the outside station, the outward bound Italian passenger liner *Toscana*, and the Queenscliff lifeboat. *Akuna* plucked twenty-one bedraggled commandos from the sea, *Toscana* picked up another eight men, and took them on to Adelaide, and the Queenscliff lifeboat rescued another five. The others managed to reach the shore safely.

Pilot Simpson, in charge of *Akuna*, was made a Member of the Order of the British Empire for 'outstanding seamanship and steadfastness in the hazardous operation.' A member of *Toscana's* crew received the British Empire Medal for gallantry.

When, in November 1961, the two thousand four hundred and fifty-nine ton *Wangara*, outward bound and exempt from pilotage, grounded near Point Lonsdale, the tug *Walumba* was sent from Melbourne to attempt to tow her off. *Walumba* was disabled and in danger of being driven ashore when the towline parted and fouled her propeller. *Wyuna* and the Queenscliff lifeboat were summoned to her assistance. *Wyuna* towed *Walumba* to safety and the pilots were awarded £10,000 salvage. *Wangara* was refloated with assistance from the tug *Geelong*.

Pilot vessels, constantly operating in close proximity to other ships, are particularly prone to minor collisions. But the nearest *Wyuna* will ever come to total disaster was while heading for Queenscliff to land pilots on 15 June 1972. Off Point Lonsdale in dense fog at 6.15 pm she collided with the outward bound four thousand one hundred and twenty-nine ton *Bass Trader*, also exempt from pilotage. Both ships met almost end on and miraculously bounced apart. Both sustained considerable above-waterline damage; *Bass Trader* had to return to Melbourne for repairs, but *Wyuna* maintained her station until the usual relief by *Akuna*.

For more than one hundred years, harbour pilots at Melbourne operated in conjunction with the sea pilots, although they held no financial interest in the pilot ships or plant. Their function was to pilot ships bound for up-river berths to and from their destination, relieving the sea pilots in Hobson's Bay; or vice-versa, for outward bound ships.

Then in 1957, protracted negotiations were concluded for the eight harbour pilots to amalgamate with the sea pilots. Harbour pilots were licensed as sea pilots and sea pilots were licensed to act as harbour pilots. To attain financial equality the former harbour pilots paid the current 'ingoing' into the sea pilots funds, their contributions giving a considerable boost to the Reserve Fund. The Salary Equalisation Fund and the Ship Replacement Fund are still in existence, but dormant.

Since 1957 outward bound ships have been piloted from River Yarra berths to sea by the same pilot. But because of the large number of ships arriving in Hobson's Bay at night time and anchoring until daylight, and to handle vessels shifting their berths, four pilots were detached from the sea roster to maintain a harbour service, all pilots taking their turn by rotation.

There are now thirty-nine Port Phillip Pilots. Vacancies are filled by certificated foreign-going master mariners with at least a year's service in command while in possession of a Port Phillip Pilotage Exemption. The age limit is forty-five and applicants are required to be in first class physical condition with a high standard of eyesight, maintained by annual examination. The list of applicants is submitted to the pilots to make their choice before the Marine Board makes a final selection.

One pilot is nominated annually to act as secretary-treasurer and is in charge of the Pilot Office at Williamstown. He is also given power-of-attorney by all the pilots individually to sign cheques on their behalf, and to handle the business affairs of the unique organisation. The secretary-treasurer is also the pilots' representative on the Marine Board of Victoria.

Seventeen of the pilots are now licenced to handle ships in Westernport, two always being available at that port. More can be detached from the sea roster if shipping movements at Westernport cannot be handled by two. Each pilot also takes a turn in charge of the cruising pilot ship, which has no permanent master. This pilot-in-charge is responsible for the safe and efficient operation of the ship, and arranges pilots for inward shipping.

Whenever possible pilots taken off outward ships are rostered for arrivals, but it often transpires that there are more departures than arrivals; surplus pilots then return to Melbourne by car or ship from the outside station. Or there may be more expected arrivals than are covered by pilots from outward bound ships. Then pilots have to go to Queenscliff by car to board the pilot ship.

At present pilots are rostered for duty for three weeks and then have a week off. When none are absent through sickness or other reasons there are seventeen available on the sea roster. They cover all shipping movements, inwards and outwards, between Melbourne, Geelong and Port Phillip Heads. There are always four on the harbour roster to berth ships that anchor overnight and for harbour removals. All of these movements are organised from the Pilot Office, situated for many years in the Customs House building in Nelson Place, Williamstown.

Two launches and a fleet of cars are also based at Williamstown, with a team of boatmen and drivers to ensure pilots are promptly delivered to ships requiring their services, or to take them off ships when their duties are concluded. Ships' agents are expected to give at least three hours notice when the services of a pilot are required.

Port Phillip Pilots operate a non-profit organisation. Pilotage fees are set by the Marine Board, which operates in a similar manner to when it was established in 1887. The Marine Board also negotiates pilots' remuneration, and has access to their accounts.

Four times a year a sub-committee assesses the financial requirements of the Pilot Service and estimates the probable financial return from the number of vessels expected. Pilotage rates are adjusted to offset one of these estimates against the other. The Marine Board also collects the pilotage fees from the shipowner or their agents. Ninety per cent of this return is paid into the account operated by the secretary-treasurer of the Pilot Service for disbursement. Six per cent is paid into the Pilots' Sickness and Superannuation Fund, and the remaining four per cent to the State Treasury for Marine Board expenses.

Since 1839 the pilots of Port Phillip have weathered many storms, physically and metaphorically. Today, as in the past 134 years, a shipmaster's welcoming handshake is often accompanied by a sigh of relief as he hands over his command to an experienced pilot who fully appreciates the captain's position.

Port Phillip Pilots have survived these years as an independent body free from industrial problems. With few exceptions ships of all shapes, sizes and nationalities have been guided safely through the notorious Rip to what is now one of the world's major ports.

This history of the Pilot Service is published in 1973 to commemorate the centenary of the dismasting of the Pilot Schooner *Rip* on 15 July 1873. This disaster to the sister-ship of the now famous schooner-yacht *America,* and the self-sacrifice of Able Seaman Marr, brought about the foundation of the Royal Humane Society of Australasia.

It is fitting that two men presently associated with the Pilot Service – former pilot Captain Sam Benson and Traffic Officer in the Williamstown Pilot Office, George Lansley – are recipients of awards from the Royal Humane Society.

During the Second World War Captain Benson was an officer in the Royal Australian Naval Reserve and in command of HMAS *Kiama.* His prompt and brave action in plunging into the sea to rescue a seaman who had fallen overboard earned him the Society's Bronze Medal.

Captain Benson joined the Pilot Service in 1949 and, after representing the Batman electorate in the Federal House of Representatives for six years, retired in 1969.

George Lansley was an able seaman in the Pilot Steamer *Victoria* when the *Lawhill,* the last large sailing ship to call at Melbourne, hove to off the cruising station on 31 May 1943. While a boat was being lowered from the *Victoria* the rope fall jammed, overturned the boat, and tipped Lansley and another able seaman, James Patrick, into the sea. Lansley kept his older shipmate afloat for nearly an hour before they were rescued, and was awarded the Royal Humane Society's Certificate of Merit for this heroic effort.

After working continuously for the Pilots in the intervening years he is now a senior boatman and in charge of traffic operations at Williamstown.

PART TWO

Port Phillip: Defences

CHAPTER 10

The Colonial Sloop 'Victoria'

In 1851, soon after Victoria gained independence from New South Wales, Lieutenant Governor La Trobe applied to the colonial secretary for a warship to guard the increasing number of ships using Port Phillip. In April 1853 the 500 ton corvette HMS *Electra*, under Captain Morris's command, was sent from Sydney. One of her first assignments was to arrest the mutinous crew of the emigrant ship *Wanata* after their seizure of a pilot boat at Port Phillip Heads. HMS *Electra* then became the headquarters of the Hobson's Bay water police.

Commissioners investigating harbour defences in 1854 also advised the Victorian government to acquire its own war steamer; as well as six nine pounder guns and two twenty-four pounder howitzers; and that the army in the colony should be augmented by a corps from India. Although the latter suggestion did not eventuate it was probably instrumental in the formation of the Volunteer Rifle Regiment, the Volunteer Artillery, the Geelong Volunteer Rifles and the Volunteer Yeomanry, under the Defence Act of 1854.

Specifications were prepared for a steam screw vessel, barque rigged, one hundred and sixty-six feet long by twenty-seven feet beam; keel of English elm, floors of English oak, diagonally planked with two thicknesses of mahogany, and decked with Danzig fir.

Victoria was the name selected for the 'Colonial Sloop'. In deference to the reigning monarch it was specified that the figurehead be 'handsome and appropriate'. The five hundred and eighty ton colonial sloop was built on the Thames by Young,

Son, Magnay and Company to drawings prepared by Oliver Lang, Master Shipwright attached to the Admiralty dockyard at Pembroke. She was pierced for six guns but only two thirty-eight pounders and one sixty-eight pounder were mounted. Captain Norman, with eight officers and one hundred and fifty men, delivered the ship to Hobson's Bay on 31 May 1856. HMS *Electra* then returned to New South Wales.

HMCS *Victoria* was classed as an armed dispatch vessel and operated under the Chief Secretary's orders to protect the interests of the public. She was kept ready for sea on 'telegraphic notice'. Her presence in Hobson's Bay with a permanent crew of fifty meant that the Water Police could be reduced in strength from forty-two to ten men. When HMCS *Victoria* had to leave port a temporary guard ship manned by volunteers took over the water police duties.

Not long after the colonial sloop's arrival doubts were raised about the vessel's authority to intercept ships on the high seas. When the American ship *General Nowell* eluded customs and police and cleared Port Phillip with writs for debt nailed to her mast, HMCS *Victoria* sailed in pursuit, caught the runaway, fired a shot across her bows when the American captain refused to heave to, and both vessels returned to port.

After the United States Consul had claimed that the colonial sloop acted illegally in intercepting an American ship on the high seas because she was not commissioned by the Admiralty, HMCS *Victoria* was never again used for a similar purpose.

During the Maori War, 1860-61, HMCS *Victoria* was temporarily attached to the Royal Navy and accompanied the 40th Regiment from Port Phillip to New Zealand. Before returning she went to the Gulf of Carpentaria to search for the missing explorers Burke and Wills, and did not resume duty in Hobson's Bay until March 1862.

When the Confederate American raider *Shenandoah* sought refuge in Port Phillip in 1865 HMCS *Victoria* was the only armed vessel in port. If the two ships had met in armed combat it was likely that the local vessel would have had a slight advantage. There were rumours that the American captain would willingly have exchanged ships. At that time a section of the

The colonial sloop Victoria, 1856-1896

Digging the moat at Fort Queenscliff

HMVS Cerberus and HMS Nelson

The torpedo-boat HMVS Childers with HMS Nelson hulked

public demanded that *Victoria* should be sold, and it was rumoured that the *Shenandoah's* captain had offered a good price for her with the intention of using the sloop as another raider.

But HMCS *Victoria* was destined to continue in the Victorian Government service. With the assistance of the locally built *Pharos* she surveyed Westernport and Bass Strait until her active career ended in 1878. Then, after being laid up for ten years, there was an unsuccessful attempt to convert her into an excursion steamer. She was finally broken up at Williamstown in 1895.

CHAPTER 11

Early Fortifications

One of the reasons Lieutenant Colonel Collins gave for abandoning his settlement at Port Phillip in 1803 was the impracticability of defending the area against attack. Marauding natives were then the threat but Captain Woodriff, in command of HMS *Calcutta,* had already reported to the Admiralty on the necessity to defend Port Jackson. During his later career in the Royal Navy, Captain Woodriff made a detailed study of colonial port defences.

Thirty soldiers attached to Captain Lonsdale, Melbourne's first Police Magistrate, arrived from Sydney in HMS *Rattlesnake* on 29 September 1836 and were the colony's first defence force. They were accommodated in wattle huts on the slopes of Burial Hill, later known as Flagstaff Hill. Seventeen years later the military headquarters of the British Army in Australia, the 12th and 40th Regiments, were transferred to Melbourne. Imperial troops and naval forces were then solely responsible for the defence of Australian colonies.

In 1858 a Royal Commission inquired into 'the naval and military means, ashore and afloat, by which an attack by an enemy may best be repelled'. In July Major General Macarthur, Chairman of Commissioners representing the Royal Engineers, the Royal Navy, the Volunteer Forces, the Inspector of Musketry, and politicians, reported:

In the event of war between Great Britain and one of the principal maritime powers it may be reasonably expected that some attempt will be made by the enemy to cripple and destroy the great and lucrative trade now subsisting between this colony and the mother country, which, in the present defenceless condition of Port Phillip, might be effectively accomplished by a single frigate, or even a privateer of a large class, sailing or steaming into Hobson's Bay, destroying or taking possession of the shipping, and thereby laying the city under contribution, without the possibility of any effective resistance being offered.

Such an enemy, the Commissioners suggested, could take up a position out of range of the guns at Williamstown and Sandridge (Port Melbourne) but among the anchored shipping, and create havoc. HMCS *Victoria* was referred to as the only effective vessel available to defend the port and she was said to be 'half armed, half manned and frequently absent'.

Only two months before the Royal Commission heard evidence, Sir John Burgoyne had recommended fortifying Port Phillip Heads. Disagreeing with this recommendation the commissioners were of the opinion that, inside the Heads was a basin of deep water where large vessels could move freely and anchor out of range of any guns then in existence that could be mounted on Point Nepean, Point Lonsdale or Shortland's Bluff. Enemy ships entering in mid channel would be within range of such guns for only a few minutes before being able to turn eastwards for South Channel and be out of effective range.

Should the lighthouses on Shortland's Bluff be operating an enemy could enter at night and anchor out of range of such guns. Batteries which necessitated the lighthouses being inoperative might cause loss of trade. Even so, with no lights to assist, an enemy could enter at dawn with English colours flying and be out of range before his true identity could be ascertained.

Forts at the Heads would not remove the necessity to defend Hobson's Bay and Geelong Harbour. Without defences nearer home the two cities would be at the mercy of such an enemy, who could secure himself against molestation on his outwards passage by demanding the evacuation of the forts as an article of capitulation.

The commissioners considered that no less than sixty guns

would suffice to effectively cover the entrance channel during the brief period an enemy would be within range. To keep sixty guns ready and manned continuously would need at least six hundred trained artillerymen, whose usefulness would cease as soon as an enemy had passed. This considerable force would then be needed urgently at Melbourne, but would have no means of getting there.

Defences at Melbourne and Geelong would not have to be continuously manned, because several hours notice would be available there if an enemy entered the Heads. Improvement of the defences on the shores of Hobson's Bay were considered to be more important than the fortifications of the Heads. However, the commissioners anticipated an improvement in the future range of artillery which would increase the time an enemy would be within range while entering. When this eventuated they recommended that the situation should be reconsidered.

Major General Macarthur's report also recommended a militia force of three thousand to four thousand in addition to regular troops, and drew attention to the fact that every country, except the Australian colonies, expected capable citizens to perform military duties. It was still on the United Kingdom statute books.

Government employees at Williamstown and elsewhere, connected with the sea, were enrolled in a naval militia and trained to operate guns ashore and afloat under commissioned officers.

When the Williamstown Marine Artillery was established by volunteers in 1856 Captain Ferguson, the harbour master, was in charge. This became the Naval Brigade in 1859, with Captain Ferguson still in charge. The Brigade's first assignment was to provide crews to man the guns of the hulk *Sir Harry Smith* which was being used as a guard ship in the absence of HMCS *Victoria*.

CHAPTER 12

Floating Forts?

The 1858 Commission's Report made the first reference to the possibility of defending Hobson's Bay with iron-plated floating batteries. At the time it was known that experiments along these lines were being carried out in England, but until the results were available no recommendations were made, except that a screw-propelled blockship should be considered. Any floating defences, it was pointed out, would be principally for the preservation of shipping, most of which was owned in Britain. For this reason the imperial government could be expected to bear half the cost.

In Britain, the Victorian government was represented by Hugh Childers, who, eight years previously, migrated to Melbourne with a Bachelor of Arts degree and a letter of introduction from Earl Grey. Although accepted in the best society his first job was as tally clerk on Cole's Wharf. Within four months however, Childers was appointed inspector of schools.

During the next two years he successively became an immigration agent; a commissioner of National Schools; director of Mount Alexander and Murray River Railway; and government auditor-general. In the latter capacity he was blamed for the deficit which led to the 1854 economic crisis. When Sir Charles Hotham became governor in 1855 Childers was collector of customs. After a Select Committee inquired into the Administration of the Customs Department, Sir Charles Hotham demanded the collector's dismissal. But influential English friends arranged for the governor to be censured instead. When elected representative government came in 1856 Hugh Childers was returned to the Legislative Assembly as the member for Portland, and nominated Commissioner of Trade and Customs.

The following year he was sent to England as Victoria's first agent-general. This taste of English politics apparently indicated that the House of Commons offered better prospects. After returning to Melbourne in 1858 he finally departed for England with a substantial colonial pension. Within two years he was representing Pontefract at Westminster and acting as honorary agent-general for Victoria.

In the early 1860s the decision not to fortify Port Phillip Heads was again reversed. Three sixty-eight pounders were installed in an open battery at Shortland's Bluff, and all the eastern shore of Hobson's Bay was proclaimed a military reserve. There were three heavy batteries between Sandridge (Port Melbourne) and St Kilda. Two suggested guard ships for Hobson's Bay did not materialise but five batteries of heavy calibre guns were installed at Williamstown.

Captain Peter Scratchley, of the Royal Engineers, was sent from England to report on the defences of the colony in 1860. Six years previously at the age of nineteen, Scratchley graduated from the Woolwich Academy. After serving in the Crimea, and the Indian Mutiny, he was promoted to captain in 1859.

Captain Scratchley recommended four heavy batteries at the Heads, and that South and West Channels be protected by forts raised from shoal water. These works would have cost £200,000 but only three-quarters of that amount was available. £50,000 had been spent when work was halted while the possibilities of floating batteries were investigated.

News had reached Melbourne from the American Civil War that a floating battery, *Monitor* mounting two guns in a heavily armoured turret, and having only two feet of freeboard, had successfully engaged the three thousand five hundred ton cut-down frigate *Merrimac,* bristling with eight guns in a huge casemate.

Most of the men-of-war built during the first half of the nineteenth century were designed for fighting at close quarters.

Guns arrayed on several decks demanded sufficient stability to limit the angle of heel created by wind pressure on sails, and to permit the tier of guns nearest the waterline to be fired. War in the Crimea proved these wooden warships ineffective against shore batteries. Floating batteries – armour plated wooden rafts known as ironclads – were improvised. A further development was Britain's first iron hulled warship, HMS *Warrior*, with a midship belt of four and a half inch thick iron, backed by eighteen inches of teak-wood. (In 1970 the Maritime Trust acquired the hull of HMS *Warrior*, still in a remarkable state of preservation, with the intention of restoring it to the original plans as an addition to the nation's maritime museums.)

With steam propulsion rapidly becoming more popular naval architecture, previously the domain of naval officers, became highly specialised. Captain Cowper Coles, the officer who had experimented with floating batteries in the Crimea was in command of the paddle steamer *Stromboli;* and towed a fortified raft close enough inshore to successfully engage shore batteries in the Sea of Azov. Public acclaim for this exploit led to Captain Coles being asked to design a fleet of similar craft for an assault on the forts of Cronstadt. Although the war ended before these craft could be tested Coles became a strong advocate of ironclads.

Encouraged by his initial success he turned his attention to designing a revolving gun turret in a ship's fore and aft line, and patented the design in 1859. Naval architects of the newer school had doubts about ironclad men-of-war fitted with revolving turrets because of their minimum fore and aft sheer and low freeboard. Furthermore sail continued to be the principal form of propulsion for ocean-going ships, steam power being an auxiliary. Masts and rigging interfered with the training of guns in revolving turrets.

Nevertheless, in 1862, Cowper Coles' design was incorporated into HMS *Prince Albert*, built for coastal patrolling. Steam was her prime mover, masts and rigging being kept to a minimum. Low freeboard and stability factors were relatively unimportant for coastal patrols. She always performed satisfactorily in her intended role.

Hugh Childers was an admirer of Cowper Coles. While *Prince Albert* was under construction he advised the Victorian government that the Secretary of State for Colonies, the Duke of Newcastle, was favourably disposed towards financial assistance from Westminster if it was decided to substitute floating batteries for fixed defences at Port Phillip Heads. Childers also told the Duke that commerce between Britain and Australia, mostly British property carried in British ships, would benefit substantially. Another suggestion by Childers was that floating batteries at Port Phillip could lessen the need for imperial troops and naval units in Victoria. In fact Childers commissioned Cowper Coles to design two iron plated vessels, each of about one thousand two hundred tons and fitted with revolving turrets. Laird Brothers, shipbuilders, had submitted estimates: a vessel two hundred and five feet long by thirty-six feet beam, with twin screw propulsion on a draught of ten feet and a speed of nine knots would cost £60,000. Alternatively, a vessel two hundred and twenty-three feet long by forty-two feet beam with twin screw propulsion on a draught of fifteen feet and a speed of eleven knots would cost £95,000. Turrets for either class would cost an additional £2000 each. One vessel could be built in ten months, the second ten to twelve months from the date of order.

If the plans were adopted another £50,000 would be needed to modify the forts in Hobson's Bay. This sum was well in excess of the budgeted £200,000 and Childers suggested that the Home government might bridge the gap. These ships were not built.

In 1864 Childers became a Junior Lord of the Admiralty in Lord Palmerston's government. Four years later he was First Lord. During this period his support was largely responsible for Cowper Coles plans being used to build the six thousand nine hundred and fifty ton ocean going turret ship HMS *Captain*. Rigging was eliminated by using tripod masts. Two three hundred pound guns were installed in revolving turrets.

After being commissioned early in 1870 HMS *Captain* displayed fast sailing qualities. With auxiliary screw propulsion speeds in excess of fourteen knots were attained. But it was this vessel's lack of stability and insufficient freeboard that caused one of Britain's worst naval disasters. On 7 September 1870 HMS *Captain* capsized in moderate weather while on a cruise in the Bay of Biscay. Her designer, Cowper Coles, with the second son of Hugh Childers and four hundred and seventy other officers and men lost their lives. There were only eighteen survivors.

CHAPTER 13

HMS Nelson

No ironclads had been ordered by the Victorian government by 1865, when the British Parliament passed a Naval Defence Act which allowed the colonies to maintain and use their own vessels of war.

It was after this Act had been passed that George Verdon, then treasurer to the Victorian government, went to England to again seek financial assistance from Britain. He was accompanied by Captain Charles Payne, chief inspector for the Victorian Steam Navigation Board.

Verdon was another English migrant turned politician. In 1851 he arrived in Melbourne at the age of seventeen and was employed in a shipping office. Not long afterwards he was in business on his own at Williamstown. Verdon's public life started as a member of the Williamstown Municipal Council. He entered the Legislative Assembly as member for Williamstown in 1859. Within a year he was treasurer for a brief period and accepted similar office under a new government in 1863. In 1866 he went to England during a parliamentary recess, and on 15 November reported that his mission had succeeded: 'The Comptroller of the Navy would arrange details of an armour-plated "monitor", or turret ship, capable of carrying twenty-two ton guns to be constructed by contract in a private yard under Admiralty superintendence'.

Verdon also succeeded in obtaining HMS *Nelson* on 'permanent loan' to the Victorian Navy.

At that time many of Britain's line-of-battle ships had been outdated in a fleet rapidly becoming steam propelled. HMS

Nelson was in this category. Although then classed as a wooden steam frigate mounting seventy guns, when her keel was laid in 1804 she was the biggest warship to be built in Britain. Commissioned in 1814 as a three-decker mounting one hundred and twenty guns she was one of the last 'wooden walls' of Britain. Sturdily built for close range battle, speed and manoeuvreability were sacrificed for size and fire-power. High sides and heavy super-structure meant that tons of ballast had to be carried to achieve the stability demanded by battle conditions. But in 1854, with other vessels of her type, HMS *Nelson* was converted into an auxiliary steamship and cut down to a two-decker.

Under Captain Payne's command she sailed from Portsmouth on 20 October 1867 and anchored in Hobson's Bay one hundred and seven days later.

Captain Payne had signed off HMS *Herald* on the Australian station in 1854 to become inspector of steam ships at Melbourne. By the time he left for England he was secretary and chief inspector for the Steam Navigation Board. After serving in HMS *Nelson* until 23 March 1870 he became chief harbour master on the death of Captain Ferguson.

On arrival at Hobson's Bay HMS *Nelson* replaced the hulk *Sir Harry Smith* as guard ship during HMCS *Victoria's* absence. Naval reservists were exercised aboard *Nelson* for the first time on 10 September 1870, when she was manned by thirty permanent staff and about one hundred Naval Brigade volunteers.

Fears of a Russian attack on Port Phillip in 1878 caused HMS *Nelson* to be further cut down to a single deck steam frigate mounting two sixty-eight pounders and twenty sixty-four pounders. A two-bladed propeller gave a maximum speed of seven knots. She remained in commission until 1891. Two years later the boilers were sold for scrap metal but there were no offers for the copper-sheathed, matured oak hull, two hundred and twenty-two feet long by fifty-six feet beam on a draught of twenty and a half feet.

The hull was eventually sold in 1898 and towed to Sydney where the superstructure was removed and the hull served as a coal hulk until 1908 when it was towed to Launceston for similar service. It was broken up on the River Derwent, in 1920.

CHAPTER 14

HMVS Cerberus

Cost of Victoria's proposed turret ship was not to exceed £125,000 of which the colonial government's contribution was £25,000 plus cost of armament. Maintenance, manning and command would be the colony of Victoria's responsibility, with occasional aid from the Royal Navy in selection of officers and men. Because the turret ship would protect British as well as colonial interests in the eventuality of war she was to be commanded by the senior officer of the station who could, in a serious emergency, withdraw her from colonial waters.

HMVS *Cerberus* was built at Jarrow-on-Tyne by Palmers Shipbuilding Company at the time when Hugh Childers was First Lord of the Admiralty. Her designer, E. J. Reed, was chief constructor for the Admiralty, one of the new school of naval architects who never approved of Cowper Coles designs. Reed and Childers disagreed over the plans of HMS *Captain*, *Cerberus* being the chief constructor's idea of a low freeboard turret ship. Four sister ships, known as the *Monster* class and designated Breastwork 'monitors' after their prototype in America, were built. Three more similar craft were built for service in India.

Reed's quarrel with Childers culminated in his resignation as chief constructor five months before *Cerberus* was launched. Childers resigned as First Lord three years later, and after the *Captain* disaster. He subsequently served as chancellor of Duchy of Lancaster; secretary of state for war, chancellor of the exchequer and as home secretary in 1886. He died at London on 29 January 1896.

He is remembered in Melbourne for the part he played in

founding the University. His first budget as auditor-general in 1852 provided £10,000 for this purpose. He was the University's first vice-chancellor and served in this capacity until 1857.

HMVS *Cerberus* was two hundred and twenty-five feet long by forty-five feet beam and displaced three thousand four hundred and eighty tons on a draught of fifteen and a half feet. There were two decks and the hull below the lower deck was divided into eight watertight compartments, all or any of which could be flooded until only the breastwork was above water. This armoured breastwork was one hundred and twelve feet long and built of four and a half inch thick Lowmoor iron. The two turrets were five and a half feet in height and armoured to a maximum thickness of ten inches. Each mounted two ten inch muzzle-loading rifled guns firing four hundred pound shells propelled by sixty pound charges. Economical speed was six knots on forty-eight tons of coal per day, maximum speed was nine knots. HMVS *Cerberus* was launched on 2 December 1868 and completed in September 1870.

Captain William Norman, who commanded HMCS *Victoria* on her delivery voyage in 1856, went to England in January 1869 to take command of *Cerberus* on completion. He died at Ramsgate eleven months later, and Lieutenant Panter, who had sailed out as navigator with Captain Payne in HMS *Nelson*, hurriedly left Melbourne in April 1870 to take command of the new ship for delivery to Port Phillip. *Cerberus* was fitting out at Chatham at the time of the *Captain* disaster.

Naval protocol decreed that *Cerberus* was not a commissioned warship until arrival at Port Phillip and she was not allowed to fly the White Ensign on the delivery voyage. Nor was there any precedent for her to fly the Red Ensign, but it was decided that she should do so and be designated a merchant ship on her delivery voyage. Because of her similarity to HMS *Captain* some difficulty was experienced in recruiting a crew of any kind. There were twenty-five volunteers aboard when she sailed from Chatham for Spithead on 29 October. Bad weather was experienced in the English Channel and *Cerberus* lumbered along with her decks awash, almost unmanageable. The volunteer crew paid off on arrival.

For the long voyage to Melbourne she was rigged as a three-masted schooner, with temporary protection around the superstructure. By the time she was ready for sea on 7 November, Lieutenant Panter had recruited a crew of sixty-five seamen.

The Bay of Biscay was in its usual turbulent mood and for four days *Cerberus* wallowed in the troughs of the seas. She would not head the wind under sail or steam. Unable to heave to she rolled to as much as 45° and both captain and crew were anxious for their safety.

But when she eventually arrived safely at Gibraltar, naval experts, recalling HMS *Captain,* tritely informed Lieutenant Panter that he should have cut away the masts to prevent her from capsizing when she rolled further than 10°!

Some of the crew refused to go any further and replacements were difficult to obtain. But the worst was over, during the remainder of the voyage, via Malta, the Suez Canal, Aden, Ceylon, Batavia and Fremantle, good weather was experienced. She arrived off Port Phillip Heads at 9 am on 9 April 1871, one hundred and twenty-three days out from England. This was her last taste of the open sea for she never again left the safety of Port Phillip.

In appearance she was unlike any warship that had previously visited Australia and her arrival in Hobson's Bay created quite a sensation. General opinion was that she was aptly named a 'monster'. Masts and rigging, and the temporary superstructure, did not enhance her ungainly lines. One newspaper's description was 'an elongated gasometer fitted with masts and sent to sea on an experimental cruise'.

For the delivery voyage she had been painted white with a yellow funnel but this was changed to traditional naval colours of the day – black hull, buff funnel and a broad royal blue band around the breastwork. Peacetime complement was twelve officers and eighty-four ratings, but there was accommodation for forty more under war conditions.

(In Greek and Roman mythology, Cerberus is a three-headed dog that guards the portals of Hades. Although the name might be apt for Port Phillip Heads its application to the Colony of Victoria is of doubtful significance.)

During fifty years of active service *Cerberus* was involved in nothing more sinister than training runs from Hobson's Bay to Queenscliff and Sorrento. Her most disastrous incident was when four men died in a torpedo explosion off Queenscliff in 1878.

These 'torpedoes' were an early form of explosive mines. Seventy pounds of gunpowder were packed into a cannister with a stick of dynamite and a quantity of guncotton. They were lowered to the seabed from boats and detonated by a trailing wire. The 1878 disaster was caused by the trailing wire fouling the boat, which was blown to smithereens with the detonation. Only one of the boat's crew of five survived.

This defensive weapon was known as a torpedo before the propelled device was invented by Whitehead in 1877; the first model, fourteen inches in diameter, was propelled at 18 knots and contained a twenty-six pound charge. Special craft, known as torpedo boats, were developed to handle Whitehead torpedoes. Two second class torpedo boats, HMVS *Nepean* and *Lonsdale,* each of twelve tons and with a speed of seventeen knots, became units of the Victorian Navy in 1884. Seven years later the seventy-five ton *Countess of Hopetoun,* a first class torpedo boat with a speed of twenty-one knots, was acquired in Britain and sailed out as a three-masted schooner in one hundred and fifty-four days. There was also a 'turnabout' torpedo launch named *Gordon.*

In 1884 two gunboats were also obtained for the defence of Port Phillip: HMVS *Victoria,* five hundred and eleven tons, one hundred and forty-five feet long and mounting a ten inch breech loader and two twelve pounders; and HMVS *Albert,* three hundred and seventy tons, one hundred and twenty feet long and mounting an eight inch breech loader, a six inch breech loader, and two nine pounders. Both vessels completed trials in England early in 1884 and sailed from Spithead on 14 February under the command of the officer-in-charge of the Victorian Navy, Captain Thomas. Another sixty ton torpedo boat, HMS *Childers,* was to join the convoy at Malta.

Under Admiralty orders for the duration of the voyage, Captain Thomas was instructed to join the Red Sea Squadron after he sailed from Malta on 26 February. On arrival at Suakim three weeks later Captain Thomas continued his voyage to Australia when he found his convoy was not needed by the Red Sea Squadron.

Leaving Aden on 31 March *Victoria* was towing *Childers*. Calls were subsequently made at Colombo, Batavia and Thursday Island. On 13 May *Childers* proceeded under her own steam when the towline parted, and the convoy entered Port Phillip on 25 June.

All of these naval units took part in what became traditional naval manoeuvres each Easter. Mock battles were staged with the forts at Port Phillip Heads and the weekend's manoeuvres concluded with an impressive searchlight display in Corio Bay.

Victoria and *Albert* were kept in commission until 1896, when the former was transferred to Western Australia, and the latter was taken over by the Public Works Department in connection with blasting operations for deepening the channel over Rip Bank. *Nepean* and *Lonsdale* were used as destroyer targets in 1912; *Childers* was hulked at Swan Island in 1918; *Countess of Hopetoun* lasted until 1924.

HMVS *Cerberus* survived. Throughout the 1914-18 war she was manned as Port Phillip's guard ship, and the magazines were used to store ammunition for the auxiliary services. When the Williamstown naval base was closed on 1 April 1921 Victoria's naval depot was transferred to Westernport, where the former South Australian cruiser *Protector* was renamed *Cerberus* to preserve the name. Renamed *Platypus II* the breastwork monitor, which had held the name *Cerberus* for fifty years, went to Geelong as depot ship for the Royal Australian Navy's flotilla of six 'J' class submarines.

She appears to have resumed her original name when sold as scrap metal in 1924 for £409. Stripped of Lowmoor iron, but retaining turrets and guns, she was acquired by the Sandringham Council in 1926. Towed to Black Rock, on the eastern shore of Port Phillip, she has since served as a breakwater for small craft.

The Cerberus Preservation Trust was established in 1970 to refloat the rusty wreck and tow it to Melbourne for preservation and possible restoration. Its historic significance goes far beyond the limits of Port Phillip.

The *Monster* class of ironclad with revolving turrets was improved upon to produce two sea-going battleships along similar lines – *Devastation* and *Thunderer* – forerunners of a new era in warship construction where main armament was contained in revolving turrets. For this reason Cowper Coles, Hugh Childers, E. J. Reed, and HMVS *Cerberus* all deserve a permanent place in the annals of naval history.

CHAPTER 15

Fixed Fortifications

Five years after the imperial troops were withdrawn from Victoria in 1870 the military defence of Port Phillip again came under scrutiny. Another Royal Commission in 1875 recommended that the strength of the militia, exclusive of the Naval Brigade, be maintained at three thousand four hundred men, two thousand garrison troops, twelve hundred infantry and two hundred sappers and signallers.

Lieutenant General Sir William Jervois was another product of Woolwich Academy, from which he was commissioned in the Royal Engineers, at the age of eighteen, in 1839. By 1855 he was in command of the Royal Engineers in the London military district. After being promoted to Lieutenant General seven years later he was appointed Director of Works and Fortifications.

In 1875, Lieutenant General Jervois was governor of Straits Settlements of Malaya and in the public eye as an advocate of oriental migration to Australia and New Zealand. It was in this capacity that he came to Australia in 1877 to survey the defence situation with Colonel Peter Scratchley. They recommended extra fortifications at Port Phillip whch eventually gave the Heads the reputation of being 'the Gibraltar of the South' – an apparently invulnerable fortress.

Fort Queenscliff became an enclosed battery mounting nine inch, ten inch and sixty pounder guns. Swan Island, covering West Channel, was similarly fortified. Two forts raised from shoals were intended to cover West, Symonds, and South Channels. Torpedo defences were also planned for West and South Channels. Then Sir William went to England, and on his return suggested modifications in the light of the increased range of modern guns.

The latest ten inch guns were reputed to be better than the older eleven inch. Longer barrels presented difficulties for muzzle-loaders in casemates and for this reason breech-loaders were recommended. Newer muzzle-loaders not in casemates could be mechanically loaded by one man, allowing others in the crew to remain behind the rampart.

By installing these improved guns Sir William believed that shore batteries could take the place of '. . . either or both of the works proposed to be erected on submarine foundations'. But after further consideration he recommended that work continue on the South Channel Fort. Deep draught ships had to use South Channel and the only site ashore that could effectively cover this channel was Point King, two miles distant from the track of ships and beyond the range of the most modern guns.

South Channel Fort was only fifteen hundred yards from a narrow portion of the channel and guns here could not fail to be effective. Iron plates in the casemate of South Channel Fort were made thick enough to withstand possible return fire from passing ships.

Popes Eye Shoal Fort was designed to protect West and Symonds Channels; Sir William Jervois was of the opinion that the newly developed 'torpedoes' would prevent an enemy from using these waterways. He considered that West Channel, which would have to be kept open for navigation by friendly vessels, could be mined with electrically detonated torpedoes which would be activated by the flick of a switch. Symonds Channel, not buoyed or usually used by shipping, could be mined to obstruct all ships, friend or foe.

When this was achieved Popes Eye Fort would not be needed to cover Symonds Channel. Swan Island Fort, which was only eighteen hundred yards from West Channel, was beyond the range of the older type guns but within effective range of the latest pieces. For these reasons work on the Popes Eye Fort was discontinued. The enclosed fort on Swan Island became a torpedo depot.

Sir William Jervois was appointed governor of South Australia in 1877 and later became governor of New Zealand from 1882 to 1889.

Another Royal Engineer, Major General Stewart, insisted that both shoal forts were still necessary and work proceeded, but was dogged by delays brought about by these top level indecisions.

A fleet of light draught schooners were chartered to carry the bluestone blocks from the quarries near Point Wilson to the South Channel and Popes Eye Forts. South Channel Fort is an artificial island raised from six feet deep water on the northern side of South Channel about three miles west of Arthur's Seat. Oval in shape, it has a maximum diameter of six hundred feet. Timber piling, closely spaced around the perimeter, allowed the bluestone blocks to be carefully placed to form a retaining wall sloping upwards and inwards to a height of about twenty feet above sea level. Contained within the reclamation are magazines and galleries of the fort. Although the guns have been removed concrete emplacements can still be seen covering South Channel.

Timber piling still binds the bluestone blocks. Since the fortifications were removed in 1914 the Ports and Harbours Department have used the magazines for storing the explosives used for blasting and extending the channel across Rip Bank.

Popes Eye Fort was intended to be a similar structure. When construction was ultimately abandoned the circle of stones was about half finished. It is now known as Popes Eye Annulus – a semi-circular stone wall projecting about eight feet from the water on the crown of Popes Eye Shoal. A beacon marking the starboard side of West Channel is erected here. On its north-eastern side the annulus is open and a theory exists that this gap was left to enable HMVS *Cerberus* to be floated within the barrier of blue-stone to serve as a portable fort. Because there is a depth of only four feet within the annulus this theory only holds this amount of water.

Swan Island Fort, with barrack accommodation and nine

'entanglements' all enclosed against assault, mounted three nine inch muzzle loaders, two ten inch muzzle loaders, and three eighty pounders.

Another suggestion for blocking South and West Channels by means of a chain boom across the fairways was rejected. But in the event of war four Melbourne owned merchant ships were selected to patrol the outer approaches to Port Phillip from which inward bound vessels could be boarded and identified. This plan developed into the Naval Control Service which operated in two world wars and for which trained personnel are available to bring it into force again at short notice.

Fort Queenscliff was encircled by a moat in 1892. Inside the moat is a brick wall, extending from Shortland's Bluff at its northern end to the original Queenscliff battery southwards. Although the drawbridge has been removed the brick structure is in its original condition, complete with loop-holes and keep. Within the fort are the lighthouses; the original Queenscliff telegraph station, which was also the Post Office; and the wooden lookout tower erected in 1850. When extra barrack accommodation was built in 1927 the lookout tower had to be raised on a brick pedestal so that a look-out could be maintained over the barrack roof.

When nine inch and ten inch guns were mounted the Low lighthouse was within the arc of fire. But Sir William Jervois did not consider this impaired the gun's efficiency because the lighthouse would only screen a passing ship from one gun at a time. He did allow the lighthouse-keepers' houses to be shifted from the line of fire.

Disagreement over fortification at the Heads persisted until 1882 when Major General Scratchley, now adviser on defences, reported that they were far from complete. The following year the Department of Defence was established with Lieutenant General Sir Edward Sargood as the first minister. Major General Scratchley became his chief adviser. Sir Peter Scratchley was appointed special commissioner for Great Britain in New Guinea in 1884. Within a year of taking up this appointment he contracted malaria and died at sea.

Fortifications at Point Nepean were similar to those at Queenscliff and Swan Island. In addition a ten inch breech-loader was mounted on Point Franklin, within Port Phillip, to cover the examination anchorage.

By 1886 the defence system was complete, including electrically operated mines in South Channel fired from a control room within South Channel Fort. Some time later a more powerful battery was mounted on Eagle's Nest, higher and eastwards from Point Nepean. On this peninsula a maze of tunnels still connects the batteries with the magazines and garrison accommodation.

CHAPTER 16

Two World Wars

By 1909 all the older guns had again been replaced by more modern pieces. Swan Island and South Channel Forts, outdated by the longer range guns at the Heads, were no longer manned. The torpedo depot had been taken over by the navy; Forts Queenscliff and Nepean were maintained throughout the two world wars.

Fort Nepean claimed the distinction of firing the first shot in the 1914 war – across the bows of the outward bound German freighter *Pfalz*. The pilot on board, unaware that war had been declared, ordered 'full astern' when the shot was fired. The captain countermanded the order but after being persuaded by the pilot that the next shot would not be across the bows he allowed her to stop. The Commonwealth government claimed the *Pfalz* as a prize of war and she continued trading under the Australian flag as the *Boorara*.

This was reputed to be the only shot fired during the 1914-18 conflict as an 'act of war' within Australian waters. Something similar developed in 1939, but the pilot had left ss *Stassfurt* when the Queenscliff gunners opened fire. The German captain refused to heave to and the ship escaped.

When the defence situation was reviewed in 1938 sporadic sea raiders were considered to be the main danger. Strong tides, and the comparatively shallow entrance to Port Phillip would deter submarine entry. Modern conditions indicated that, although cruisers or armed merchantmen might attempt bombardment they were unlikely to make a hostile entry. Although motor torpedo boats might attack shipping in Bass Strait, within

Port Phillip was considered relatively safe and booms and defensive mining were not considered necessary.

Probably the nearest the port came to attack was during the cruise of the German raider *Pinguin* in 1941. This former seven thousand seven hundred and sixty-six ton merchant ship was armed with six 5.9 inch guns, four torpedo tubes and carried three hundred mines when she left Germany. On 7 October, after a devastating cruise in the Atlantic and Indian Oceans, *Pinguin* captured the Melbourne-bound tanker *Storstad* off Western Australia. Requisitioned as a minelayer and renamed *Passat,* the tanker took one hundred of the *Pinguin's* mines. While the raider proceeded to mine the approaches to Sydney, Newcastle, Hobart and Adelaide, the tanker disposed of her quota by mining Bass Strait; at one time operating within signalling distance of the examination steamer off Port Phillip Heads.

Appraisement immediately prior to Pearl Harbour indicated that Melbourne's naval and air reconnaissance were inadequate. There were seven outdated guns only at the Heads and no anti-aircraft defences. It was estimated that three squadrons of carrier-borne dive bombers could have effectively silenced the forts and left the port open to invasion. Naval and military authorities decided that, in the event of attack, mines and block-ships would be used to close the channels. Because no mines could be spared in the invasion scare after Pearl Harbour, hasty plans were made to scuttle ships in the fairways in case of emergency. Guns were also so scarce that one of the Queenscliff pieces was taken to defend Newcastle Harbour in New South Wales.

Another defence proposal was for torpedo tubes to be mounted on two Portsea jetties, just within Port Phillip Heads. But the jetties were found to be too short and unstable. Instead a modern innovation known as an 'electric eye' was installed near the Popes Eye Shoal, supposed to be capable of giving an alarm if the beam was broken by a passing ship. This proved unsatisfactory because small craft, and certainly submarines' periscopes, would not trigger the alarm. Nor was it effective in fog, heavy rain, or smoke.

Towards the end of World War II plans for dual purpose guns to be mounted on Point Nepean and Lonsdale as a defence against small craft and aeroplanes were also abandoned. Then, in September 1944, with the Japanese being repulsed and the Germans on the brink of surrender, Melbourne, Adelaide and Hobart were no longer considered 'defended ports'.

Soon after the end of the war coastal defences generally were considered to be outmoded. Fort Queenscliff's guns were removed in 1946 and the premises have since been occupied by the Australian Staff College. New buildings have been added but all the old ones are preserved intact.

Similarly, four hundred and twenty acres of military reserve on Point Nepean, although unfortified, is still under army occupation. Since 1951 it has been attached to the Army Officers Cadet College in the old quarantine station at Portsea. And the seven hundred and seventy-six acres of quarantine area controlled by the Health Department is similarly used. Although the Health Department's buildings are maintained in a state of readiness should they be required to accept a shipload of quarantined passengers, the advance of air travel renders this more and more unlikely.

There is a move afoot to have this area declared a National Park.

Many years ago the effective fortification of the Port Phillip Heads rendered the Hobson's Bay batteries needless. The military reserve that existed between Port Melbourne and St Kilda is the esplanade known as Beaconsfield Parade, with no visible signs of its former purpose. There is evidence of the former fortifications at Williamstown, where a cricket ground has been laid out over some of the underground magazines and enclosing a portion of the former defence reserve.

Pilots, lighthouses, and defences are an integral part of port development. Shortland's Bluff, and Queenscliff, have been a focal point of the three services.

Fortifications are no longer deemed to be necessary but the

army still occupies Fort Queenscliff, where the Australian Staff College provides senior officers with advanced tactical training. Within the fort's encircling brick wall – keep and loopholes are intact but the moat has been filled in – big gun emplacements and magazines bear witness to its former significance.

All the lighthouses for the entrance fairway's complicated system of leading lights are also within the fort, making the area rich in historical significance.

For 134 years Port Phillip Pilots have also used Shortland's Bluff as a base for operations. The trim lines of the *Wyuna* are familiar to holidaymakers at Queenscliff and Point Lonsdale as, day and night, pilots board inward ships well clear of the Rip's notorious dangers or leave outward ships that have cleared those dangers.

While the history of Port Phillip defences is now a closed book that of the pilots is still being written. Important changes are in the offing but the succession of master mariners who have built this unique and efficient organisation will ensure that its high world-wide reputation will ever be maintained.

List of Pilots

The name Liley has appeared several times among the Port Phillip Pilots during the past one hundred and twenty years. William Thomas Liley was the first: he joined the service on 4 April 1857, and resigned in 1879. Prior to his arrival at Port Phillip, in the *Gloriana* in 1852, William Thomas served with Trinity House in England and was at the Farne Islands when the immortal Grace Darling figured prominently in the rescue of nine survivors from the wreck of the *Forfarshire* in 1838.

Between the time he arrived at Melbourne and his appointment as a Port Phillip Pilot, Captain Liley was in command of River Murray steamers. Soon after he retired, Thomas James Liley (junior) was granted a Port Phillip Pilot's licence at the age of twenty-eight, and retired after thirty-four years service in 1913. His father, Thomas James Liley (senior) became a harbour pilot at the age of forty-nine in 1882 and retired in 1899.

Another William Thomas Liley joined the sea pilots in 1892 at the age of thirty-five and retired in 1922. William Thomas Gordon Liley was licensed as a sea pilot from 1919 until he retired in 1948, and a relative is presently in the service: Richard William Liley, licensed for Port Phillip and Westernport, joined the service on 10 October 1965 at the age of thirty-seven.

	Age at entry	Birthplace	In Pilot Service
George Tobin	32	England	1839 - 1853
Thomas H. Sutton	45	England	1839 1853
Alexander McPherson			1840
William Timothy	31	England	1841

	Age at entry	Birthplace	In Pilot Service	
Thomas McClelland			1841	
Edwin W. Toby			1841	
James Jacks			1842	
Thomas Ruffle (or Russell)			1842	
Julius B. Thompson			1842	
John Henry			1842	
James Welsh			1846	1855
Henry Taylor			1851	1858
Arnold Grainger			1851	
Thomas Robertson	47		1851	1878
John Steel			1851	
Charles Ashby			1851	
James Cordell			1851	1866
James G. Caught	23	England	1851	1880
Daniel Organ	25	England	1851	1880
Henry J. M. Draper	25	England	1852	1889
William McQueen	36	Scotland	1852	1874
John Martin			1852	
Charles Kennedy		Scotland	1852	1876
James R. Rich		England	1852	1879
David Kerr			1853	1872
George Mansfield			1853	
John Nicholson	26	Scotland	1853	1878
George Wray			1853	
John Stewart			1853	
Thomas Hannah			1853	1874
Jennings W. Singleton	28		1853	1883
Robert Mills			1853	
William Nicholson		England	1853	1865
James P. Best	32		1853	1881
Samuel Evans			1853	1889
Hugh Williams			1853	1868
James Ridley			1853	1866
John G. Leighton			1853	1874
George M. Bower	35		1853	1884
Henry Reed	25		1853	1884
James Deane			1853	1866

List of Pilots

	Age at entry	Birthplace	In Pilot Service	
F. W. Deane	34		1853	1889
William Neil (or O'Neil)	25		1853	1881
William Rockett			1853	1876
Andrew Wauchope	23		1853	1883
William Jones			1854	1865
James Gaunson	39		1854	1870
William Hansen			1854	1871
J. Louttit			1854*	
George Dix			1854	
F. Paterson			1854	
William T. Liley	39	England	1857	1879*
William Murrell	28	England	1857	1894*
John Poole	50	England	1859	1868*
George C. Cowley		England	1859	1871*
John B. Tregurtha	40	England	1859	1888*
William McLean			1863	
John McKenzie	42	Scotland	1871	1873
William Paxton		Scotland	1873	1894
Thomas Gafford	30	England	1874	1893
Amadeo Loiseau	36	England	1874	1883
James H. Barber		England	1875	1898
Thomas G. Seal		Tasmania	1875	1907
Charles B. Blanchard	31	England	1876	1910
Janus B. Maynard	34	USA	1876	1893
Charles R. Devlin	36	NSW	1877	1891
George F. Dennis	32	England	1878	1904
George Bevis	33	England	1879	1890
Richard Dawes	32	England	1879	1911
Thomas J. Liley Jnr	28	England	1879	1913*
John F. Nelson	38	England	1880	1889
John Cairns	37	Scotland	1881	1889
John L. Stalker	41	England	1881	1905
Edward Thurnall	28	England	1881	1892
Thomas J. Liley Snr.	49	England	1882	1899*
Henry Press	39	England	1883	1909
John Stewart	36	Scotland	1883	1910

* Harbour Pilot

	Age at entry	Birthplace	In Pilot Service	
Edward Clayton	34	England	1884	
Francis C. Burton	39	South Africa	1884	1892
Alexander McWilliam	34	Ireland	1884	1914
Alexander McIntosh	33	Scotland	1884	1899
John Craig	40	Scotland	1885	1907*
George Service	40	Scotland	1885	1904
Charles Ashford	41	Scilly Isles	1885	1900*
James A. Mitchell	31	Scotland	1885	1900
William H. Emmerson	34	England	1886	1909
Richard Wills	37	England	1886	1910
Daniel Russell	36	Scotland	1886	1909
John Bennett	37	Scotland	1888	1899
William Dow	36	Scotland	1889	1919
Louis J. D. Schutt	41	Heligoland	1889	1910
Frederick W. Strickland	37	England	1889	1916
Frederick H. Stanford	38	England	1889	1909
David Doughton	38	Wales	1890	1917
Lewis B. Campbell	38	Scotland	1892	1907
William T. Liley	35	England	1892	1922
Holdsworth Nicholson	38	England	1892	1903
Arthur F. Anthony	38	England	1900	1919
Alfred Easton	39	Scotland	1900	1924
George T. Rose	41	England	1902	1926
Walter McD. Hipgrave	43	England	1902	1924
Clifton W. Palmer	31	England	1905	1939
James Rattey	31	England	1906	1939
James Smart	36	England	1907	1935*
J. H. Fearon		Victoria	1907	1928*
James McFerran	40	Ireland	1907	1928
Henry J. Blanchard	45	NZ	1907	1917
Montgomery Robinson	40	Victoria	1907	1932
Harold C. W. Thomsett	36	England	1908	1936
Alexander M. Brebner	37	Scotland	1909	1931
James Banks	36	Scotland	1909	1935
William E. G. C. Howard	36	England	1909	1938
John Parry	40	Wales	1910	1934

* Harbour Pilot

List of Pilots

	Age at entry	Birthplace	In Pilot Service	
John Paterson	38	Scotland	1910	1928
Henry F. Joynt	40	NZ	1910	1926
Ernest R. Stewart	35	Victoria	1910	1939
Reginald L. Instone	33	NZ	1910	1941
William Foulis	35	England	1911	1940
Donald K. Urquhart	36	NZ	1914	1943
Alexander Campbell	39	England	1916	1942
Frederick G. Kendall	39	Victoria	1917	1925
William T. Gordon Liley	33	Victoria	1919	1949
James A. Robilliard	40	Victoria	1920	1941
John S. Stuart	36	NZ	1921	1946
Peter Low	37	Scotland	1921	1948*
Forbes Keith A.	38	NSW	1923	1948*
Garnet C. Dalziel	41	Scotland	1925	1943
William E. Eglen	40	England	1925	1950
James G. S. Doorly	45	West Indies	1925	1945
Hereward W. Bold	44	NZ	1926	1947
John E. Walker	35	NZ	1927	1957
William H. Davies	36		1928	1939*
William Chapman	37	Tasmania	1928	1952*
Frederick C. L. Dunn	39	NSW	1929	1955
George A. Molyneux	38	England	1935	1962†
Ernest A. Stein	44	NZ	1936	1956
Cedric J. Hackworth	44	NZ	1936	1957
Robert B. Denniston	40	NZ	1936	1960
Charles R. Dingle	41	Sth Aus	1938	1959
Andrew R. Jackson	38	Ireland	1939	1962
John D. Lundie	43	NZ	1939	1943
George H. Martin	43	NZ	1940	1960
William G. Balharrie	43	Scotland	1943	1965
Robert N. Turner	43	NZ	1943	1958
Allen Murray	42	Scotland	1944	1958†
Maurice R. Tregale	48	England	1944	1959
Sidney Chaffers	46	England	1944	1962
Roderick A. Fraser	37	Scotland	1945	1972

* Harbour Pilot
† Harbour Pilot until 1957

	Age at entry	Birthplace	In Pilot Service	
Colin E. Goss	44	Victoria	1947	1967
Jack O. V. Young	45	England	1947	1966
James Roberts	45	Scotland	1948	1966
Basil Paul	43	England	1948	1969†
William R. Fellowes	47	Victoria	1948	1950
Ian Bruce Dymock	48	NZ	1948	1950
George S. Duckett	47	England	1948	1965†
Samuel J. Benson (MHR 1963-69)	39	Sth Aus	1949	1969
Roderick K. MacKenzie	38	NZ	1949	1966
Francis W. Reid	37	Scotland	1949	1969
William H. McG. Boyle	40	NSW	1950	
Charles A. Whyte	40	NSW	1950‡	
Eric A. Corker	35	England	1950	
John A. MacRae	34	NSW	1950†	
Alfred Goodale	37	NZ	1950†‡	
John H. Bell	39	Scotland	1952†	
George A. Simpson	37	Scotland	1953	1969
Robert W. Hansen	42	Victoria	1955‡	
Norman E. Farrell	42	England	1956	1961
John C. Chapman	34	Victoria	1956	
Frederick S. Angelin	38	Sth Aus	1957	
John R. Barclay	45	Queensland	1959‡	
Ian J. Sladen	37	England	1959‡	
John A. Noble	45	England	1959	
Colin S. Randall	34	NSW	1959‡	
Richard H. Clay	42	Victoria	1960	
Clive C. Springall	34	Victoria	1960‡	
Patrick J. G. Hearn	35	Sth Africa	1961‡	
Douglas J. McDonald	40	NSW	1961	1972‡
Alexander B. Mac Adie	38	Victoria	1961	
John G. V. Abbott	42	Victoria	1961‡	
Derek C. Howard	37	Victoria	1961	1971‡
John J. Tayler	33	NSW	1962‡	
James H. Nicolson	40	Scotland	1962	

† Harbour Pilot until 1957
‡ Also Westernport Pilot

List of Pilots

	Age at entry	Birthplace	In Pilot Service
Lachlan A. MacSween	36	Victoria	1962‡
Robin W. Fisher	34	Sth Africa	1963
Keith Elford	37	NSW	1965
John G. Taylor	35	Queensland	1965
Richard W. Liley	37	England	1965‡
Keith R. Jordan	38	NSW	1965
Eric M. Trower	41	Queensland	1966‡
Robert S. Fernon	38	NSW	1966‡
Brian W. T. Lewis	35	Victoria	1966‡
Howard L. Tyler	39	Victoria	1967
Norman Johnson	38	Wales	1969
John R. Nance	41	Victoria	1969
Kenneth H. Ross	33	NZ	1970
Frank N. Jolly	43	Victoria	1971
Peter K. Phoenix	41	England	1972
Neville G. R. Daniel	33	Victoria	1972
John Craig	44	England	1972
Francis D. Ingram	44	Scotland	1972
Ian R. McCaskill	44	NZ	1973

‡ Also Westernport Pilot

Some information, particularly about early pilots, may have been omitted or may be incorrect. Any reader in possession of further information or particulars is invited to forward these to the Pilot Office, P.O. Box 46, Williamstown, Victoria, 3016.

Index

Index

Index